STATISTICS
FOR
PROPERTY PEOPLE

by
DAVID B. EDELMAN

Bank of Scotland
Formerly Senior Lecturer in Statistics in the Department of Mathematics
and Statistics, Paisley College of Technology

1986

THE ESTATES GAZETTE LIMITED
151 WARDOUR STREET, LONDON W1V 4BN

First Published 1986

ISBN 0 7282 0092 9

Printed in Great Britain at The Bath Press, Avon

Dedication

To Susan, without whom . . .

Acknowledgements

I am grateful to the following:

to the Literary Executor of the late Sir Ronald A. Fisher, F.R.S. to Dr Frank Yates, F.R.S. and to Longman Group Ltd, London for permission to reprint Tables IIi, IV, and V from their book "Statistical Tables for Biological, Agricultural and Medical Research" (6th Edition, 1974)

to the Crown for kind permission to use data originally published by the Department of the Environment in "Housing and Construction Statistics" and to use data originally published by H.M. Customs and Excise in "Statistics of Trade Through U.K. Ports"

to Waddingtons Games Ltd. for kind permission to reproduce Monopoly money

to the International Monetary Fund for kind permission to use data originally published in "International Financial Statistics"

to George Outram & Company Ltd for kind permission to use data originally published in the "Glasgow Herald"

to The Building Societies Assocation for kind permission to use data originally published in the "BSA Bulletin"

to the Nationwide Building Society for kind permission to use data originally published in their "Bulletin"

to the Bureau of Labor Statistics for kind permission to use data originally published in the Economic Report of the President

to Penguin Books Ltd. for kind permission to reprint from "Facts in Focus" by the Central Statistical Office (Penguin Books, 1974) copyright (c) Central Statistical Office 1974, 1980, Chart 32 (on page 185), Chart 8 (on page 70), Chart 14 (on page 96) and half of Chart 21 (on page 131) as Figures 3.8, 3.9, 3.10 and 3,3, respectively.

Contents

viii   *Contents*

PART SIX
Applied Problems

Preface

In writing this book, I looked at many other books on Statistics and found much at fault. Many of them were difficult to follow. Many of them were too general in application. Many of them only presented material which appears in the syllabus for a particular professional exam. Many of them were static and taught the reader the techniques used at present but omitted the techniques which would be used in five to ten years' time. Many of them tried to do too much.

I have been aware of these faults and have tried not to commit them in this book. I have tried to make the book easier to follow and it is presented in sections, as well as in chapters. Thus, it will be easier to locate topics or to omit topics as required. It will also be easier to understand where a topic fits into the subject of statistics.

The contexts and examples in the book are geared towards students and practitioners in the land professions. Thus, the reader should feel some familiarity with the terms and problems encountered.

The material is not that prescribed on a particular professional or academic syllabus. It is material which is useful to the practitioner together with a fair degree of explanation so that a non-technical person can get a reasonable idea of what is being done.

The techniques presented are not merely those being used at present. Rather, there is a mixture of present-day techniques and those techniques which I feel will be adopted in the near future by the professions. Thus, the student who understands all these techniques will not only be able to perform a worthwhile function in a professional organisation but will also be better equipped for future personal advancement.

It is true that many books try to do too much. I have taken the view that this book is not intended to teach you all you need to know about statistics. For most professional people, statistics is a subject which is to be tolerated. A book will not bring the subject to life or make it exciting. Only a good teacher can do that. Thus, this book should be used as a handbook of statistics to supplement the material, explanations and examples presented in a formal course of instruction.

As was stated above, the text is laid out in sections as well as chapters. The first section is the Introduction and consists only of Chapter 1. Chapters 2 and 3 make up the section on Displays of Data. Chapters 4 and 5 comprise the section entitled The Basic Language. Chapters 6 to 9 deal with the basic theoretical tool of the statistician. Probability. Chapters 10–13 comprise a section entitled Pure Problems while Chapters 14–17 deal with Applied Problems. Thus, it is hoped that this will clarify matters and enable the reader who wishes to become acquainted with only a small area of statistics to select those sections and chapters relevant to his/her needs.

In writing this textbook, I am indebted to many parties. I firstly must thank my teachers for their skill and wisdom (and patience). I must also acknowledge a debt to hundreds of textbooks which have passed through my hands. It is impossible to say which texts are responsible for my understanding and the approaches I take; my debt is to all the authors concerned.

I must thank Professor Alan Millington for helping me to germinate the seeds of this text and, of course, the Estates Gazette Ltd for producing the fruit.

Thanks, of course, go to my colleagues in the Statistics group of the Department of Mathematics and Statistics at Paisley College of Technology. The many discussions and arguments we have only serve to improve all our teaching ideas.

PART ONE

Introduction

CHAPTER 1

Introduction

1.1 What is Statistics?

The Financial Times Index fell by 5.1 points yesterday.
House Prices have risen by 2.7% in the last quarter.
Managers in Austria earn 16% more than managers in France.

Everyday we are faced with such statistical findings but how have they been reached and what is their applicability? These are only a few examples but you should be able to think of many more. These are one type of example of the produce of statistics for, very broadly speaking, we can define statistics to be the science of collecting and analysing information for the purpose of reaching conclusions and making decisions from situations of uncertainty. In addition, a good statistician will present the information and the conclusions in a manner which is easily understood by a non-specialist.

1.2 What is Statistical Inference?

Most information with which we deal is numerical information or data. Thus, we are not usually interested in the fact that 12 Elm Avenue is a semi-detached villa. Rather, we are interested in the fact that there are 14 semi-detached villas, 6 detached villas and 22 bungalows in Elm Avenue.

In examining the housing stock in a town, for example, we rarely look at every residence. It is too time-consuming and costly. We refer to all the houses in this town as the population and usually we take a sample from this population. We may be interested in the average value of the housing stock for this town. Such a measure of a characteristic of the population is called a parameter. A corresponding measure which is used only to describe a sample is called a statistic.

The main purpose of statistical analysis is to use sample statistics to estimate population parameters. This is called statistical inference.

1.3 Types of Data

1.3.1 Quantitative and Qualitative Data

The first distinction that needs to be drawn is the different types of answers to "What colour of hair do you have?", "Do you like Beethoven?", "What time did you have breakfast?", and "How old are you?". The first two questions give rise to qualitative data while the last two give rise to quantitative data. We often refer to the answer of such a question as a variable or random variable. The basic difference is that qualitative data are categorised—we have several categories of hair colour but we cannot place them in order or assign numerical values—while quantitative data are numerical. Note that sometimes we deal with ranks. Here, the value of the variable may be, for example, 1, 2, 3 or 4 depending on whether this value was the smallest of 4, the second smallest, the second largest or the largest. In this case, the values ranked may be qualitative or quantitative. The ranks, however, appear quantitative but great care must be taken in the analysis.

1.3.2 Discrete and Continuous Data

In this text, we deal almost exclusively with quantitative data but other distinctions must be drawn. Data may be either discrete or continuous. One way of defining this is to pick a value of random variable and ask yourself what the next highest value is. If you can say what it is, then the variable is discrete, otherwise it is continuous. For example, if the variable was number of children in a family, the next highest number after 4 is 5. Therefore, this is a discrete variable. On the other hand, if the variable was temperature and the temperature was 22.6°C, the next highest temperature is 22.6000...0001°C where there are an infinite number of zeros. Thus, we can say that, in general, discrete random variables arise from counting something while continuous random variables arise from measuring something. The confusion occurs as we generally report continuous random variables in discrete units. For example, we may report someone's height to the nearest half inch or centimetre. However, if we have eight men whose heights are reported to be 1.79 m, it is certainly not the case that they are all of exactly the same height.

1.3.3 Primary and Secondary Data

As we stated above, statistics mainly uses data collected in a sample. This is done as it cuts down on time and cost and is less cumbersome.

However, it is cheap and quick to use data that someone else has collected or to use published data such as government statistics. If data are collected specifically for the purpose of analysing the question under consideration, the data are called primary data. If, on the other hand, the data are collected originally for some other purpose and we are now using them, they are called secondary data.

There are advantages and disadvantages attached to each type of data. These we shall not discuss in depth. The main disadvantage with secondary data, however, is that they are not our data and, therefore, we do not know exactly how they are collected or where, or exactly how the variables are defined. For these reasons, while we use secondary data frequently, we must be wary of placing too much confidence in them.

1.4 Some Omissions

After finishing a traditional statistics course, one of the most enduring memories for the majority of students is the t-test. This is a procedure commonly used when the sample is small and little is known about the value of the population parameters. However, this procedure is difficult for some students to follow and, provided we have samples of sufficient size, we can use simpler procedures which are almost equivalent. Therefore, the decision has been taken to omit t-tests.

Also, categorical or qualitative data are often analysed in a particular way called contingency table analysis. This is, in reality, a very large topic which it is better to ignore in this text. Thus, contingency table analysis is omitted.

1.5 Conclusion

In concluding this introduction may I wish you luck. Statistics is a subtle blend of common sense and mathematics, and both of these are difficult enough to grasp on their own. However, as an applied subject, it has its own rewards as the mist clears over the mountain of data and an answer becomes visible.

Take care as you proceed and be alert. Don't become just another statistic; aspire to becoming a statistician.

PART TWO

Displays of Data

CHAPTER 2

Tabulation

2.1 Introduction

As a means of presenting data, tabulation is simple yet powerful. While it appears easy to construct a table, there are many points that we can learn and adopt as "good practice". These differentiate between useful tables and useless tables and we shall try to cover these in this chapter.

Before we proceed, let me give you a warning. Most of this chapter should be common sense. As a result, you may be saying to yourself "That's obvious" or "Of course" or "I don't need a book to tell me that". However, what we shall try to do is to instill these ideas not into your mind but into what you actually do in practice.

2.2 An Example of Tabulation

Table 2.1 Size of Workforce and Average Weekly Earnings in USA, 1975

Type of Employment	Size of Workforce* ('000s)	Average Weekly Earnings
Manufacturing	18,347	189.51
Contract Construction	3,457	265.35
Wholesale and Retail Trade†	16,947	108.22

Source: Economic Report of the President, 1977.
* Size of Workforce is the number of wage and salary workers in non-agricultural establishments
† Includes restaurants, bars, cafés, canteens, etc.

Is this table useful? As a means of conveying information, the answer must be in the affirmative. The main reasons are brevity and clarity. The table conveys the information far more quickly than words could and it is far easier to see what the information is and to extract one or two numbers if required.

Consider the table below. Such a table is concise and clear. We can develop the table by calculating, for each age group, the percen-

Table 2.2 Distribution of Personal Income before Tax and of Age*, UK 1982

Income (£)	Age (Years)					
	16–25	26–35	36–45	46–55	56–65	66 and over
0– 3999	531	139	64	37	85	139
4000– 6999	1814	1435	243	86	161	226
7000– 9999	1305	3466	2831	936	831	539
10000–12999	106	2713	3053	1391	1283	214
13000–15999	22	311	1268	1482	1504	62
16000–19999	1	29	133	721	631	12
20000–24999	0	2	11	103	70	0
25000 and over	0	0	1	5	4	0
Number of people	3779	8095	7604	4761	4569	1192

Source: Hypothetical Data
* Data are a sample of those in full-time or part-time employment in an English town.

Table 2.3 Percentage form of Table 2.2. Percentages by age band*

Income (£)	16–25	26–35	36–45	46–55	56–65	66 and over
0– 3999	14.1	1.7	0.8	0.8	1.9	11.7
4000– 6999	48.0	17.7	3.2	1.8	3.5	19.0
7000– 9999	34.5	42.8	37.2	19.7	18.2	45.2
10000–12999	2.8	33.5	40.1	29.2	28.1	18.0
13000–15999	0.6	3.8	16.7	31.1	32.9	5.2
16000–19999	†	0.4	1.7	15.1	13.8	1.0
20000–24999	0	†	0.1	2.2	1.5	0
25000 and over	0	0	†	0.1	0.1	0
Number of people	3779	8095	7604	4761	4569	1192

Source: Hypothetical Data
* Percentages are quoted to the nearest 0.1. As a result, the percentage in a column may not add up to exactly 100.
† This denotes that some number was recorded which was less than 0.05%.

tage falling in each bracket. However, we ought to note that the income brackets are not of equal width.

Having derived a percentage table we are now able to compare income distributions of different ages. However, we have kept the totals in each age group in the table and we could, therefore, derive Table 2.2 from Table 2.3 with only small errors.

Table 2.4   Type and Size of House Advertised*, 7 August 1984

Type of House	No of Bedrooms	Facilities A	B	C	D
Villa	1	0	1	0	0
	2	4	4	0	0
	3	3	10	2	3
	4	2	3	1	4
Flat	1	8	1	1	0
	2	7	3	0	1
	3	2	3	1	3
	4	0	1	0	0
Bungalow	1	0	0	0	0
	2	4	4	1	0
	3	2	9	2	3
	4	0	1	2	4

Source: Glasogw Herald, August 7, 1984
* A selection of 100 advertised properties was made. The properties were placed in one of three broad categories: Villa, Flat and Bungalow. Further, the facilities were classified as:
A: one toilet, no central heating
B: one toilet, central heating
C: more than one toilet, no central heating
D: more than one toilet, central heating

The figures appearing are the frequencies of occurrence of that particular classification of property, e.g. there were 10 properties which were 3-bedroomed villas, centrally heated with one toilet.

This table is different from the three before in that more than two pieces of information are recorded on each observation, i.e. on each property. Another difference is that two of the variables on which information was collected viz. type of house and the existence of central heating, are nominal variables. In statistics, this type of table in often called a contingency table. There are many methods for analysing these tables which are well covered in other textbooks and will not be discussed here.

However, note that, if desired, we can simplify the table. Suppose, for example, we are interested in the relationship between facilities and number of bedrooms, irrespective of the type of house. We can simplify the table by "collapsing over house type".

We merely take the figures for the different types of house and add them up. For example, if we wish to know how many 3-bedroomed properties had one toilet and had central heating, we add the 10 villas to the 3 flats to the 9 bungalows to get 22 properties. The complete table is Table 2.5.

Table 2.5 Size and Facilities of Advertised Properties*

Number of Bedrooms	Facilities			
	A	B	C	D
1	8	2	1	0
2	15	11	1	1
3	7	22	5	9
4	2	5	3	8

Source: Glasgow Herald, August 7, 1984
* For explanation of terms and definitions, see Table 2.4

If we are only interested in toilets and bedrooms, we can collapse further to Table 2.6

Table 2.6 Bedrooms and Toilet Facilities of Advertised Properties

Number of Bedrooms	One Toilet	More than one Toilet
1	10	1
2	26	2
3	29	14
4	7	11

Source: Glasgow Herald, August 7, 1984

If we assume that properties have either one or two toilets, then we have 100 properties with 128 toilets and 177 bedrooms, i.e. on average these properties have one toilet for every 1.38 bedrooms.

2.3 Validity

We must be very careful about whether the data presented in a table are primary data or secondary data. In particular, it is very poor practice to draw conclusions using secondary data without first examining their source. In the data in Tables 2.4–2.6, we may wish to ask if the data are primary or secondary. If they are secondary, then we will want to know the purpose for which they were collected. We will also want to know if the day on which they were collected was an "average" or "ordinary" day for property advertisements. We must also consider how sensitive our conclusions might be to the definition of villa, flat and bungalow.

Do not feel that we can never reach a conclusion as we can

always bring up other questions and objections. However, it is important to understand the limitations of our data and of our conclusions.

2.4 Tabulation—Good Practice

The following is a list of points of good practice with comments and references where appropriate. A good table should have:

A title: Where it is not obvious, the title should include the date when the data were current. If any terms in the title need explanation, this should be carried out in a footnote (VID Table 2.4).

Labels: It should be clear what each row and column in the table is. In addition, the units of measurement should be included (VID Table 2.2).

Useful Calculations: If they will be useful and unambiguous, percentages, tables, etc. should be included. However, in Table 2.4, I felt that totals may be ambiguous as it may be unclear as to exactly what was being totalled.

A Source: Wherever possible, include a source. This will enable a reader to check the data and also to decide if these data can be used as secondary data to answer his/her questions.

In addition, wherever it will be of use, tables should have footnotes explaining the meaning of the entries or of the labels (VID Table 2.1, 2.3) and have warnings which may resolve anomalies in the data (VID Table 2.3).

2.5 A Warning

Having discussed points of good practice, I cannot promise always to follow them in the rest of this text. However, a more important point is that you should not go away with the idea that these are hard and fast rules. They are guidelines which should encourage you to think and to construct better tables of data in various contexts. Use them thoughtfully and wisely.

CHAPTER 3

Pictures

3.1 Introduction

In this chapter, we shall examine various ways of pictorially presenting data. Some of this chapter should be common sense—the more the better—but it is still necessary to cover the basic ideas and advisable to list and discuss various types of pictorial representation.

The methods of presenting data below will have advantages and disadvantages. As a result, some will be of good for some uses and some for others. Hopefully, the use for which a type of presentation will be good or bad will be clear. The important guideline here is that a pictorial representation should be clear and it should be useful. If either of these is not true, do not bother to construct the picture.

3.2 Line Graphs

A line graph is often used to pictorially represent the change in a factor or variable over time. For example, we could be interested in the Gross National Product of a country for the period from 1905 to 1955. This could be shown as in Figure 3.1. It could be embellished as in Figure 3.2, although this may lessen the impact of the data.

We are not restricted to having only one set of data on a graph. It is possible to have, for example, plots of the hours of sunshine at two weather stations on the same graph. We can even put sets of data with different scales/units on the same graph. For example, we could produce a graph of a country's population over a 10-year period and the amount of government subsidy to the arts. These will be measured in different units and the numbers are likely to be of disparate size. However, for ease of comparison, we would want the plotted lines to be about the same place on the actual graph. Therefore, all that is needed is to have two vertical scales, one on each side of the graph, one for each of the two pieces of information which are plotted on the graph.

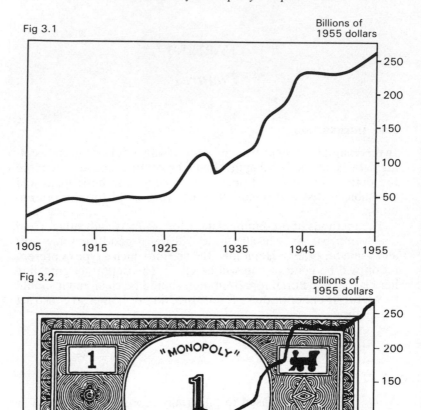

Fig 3.1

Billions of
1955 dollars

250

200

150

100

50

1905   1915   1925   1935   1945   1955

Fig 3.2

Billions of
1955 dollars

250

200

150

100

50

1905   1915   1925   1935   1945   1955

In Figure 3.3, we have done this but for data relating to cinema admissions and television licences.

Obviously, provided we can solve the problem of different units/scale, we can plot three or even more pieces of information on the same graph. For example, in Figure 3.4, we have plotted values of new contractors' orders for three regions.

A line graph is very often used to represent the change in one or more variables with time. As a result, it is sometimes referred to as a Historigram.

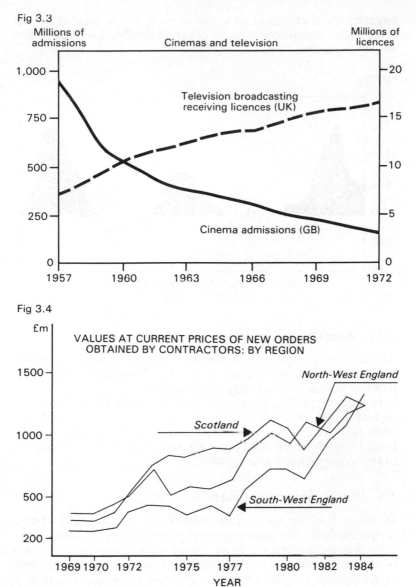

Fig 3.3

Millions of admissions — Cinemas and television — Millions of licences

Television broadcasting receiving licences (UK)

Cinema admissions (GB)

Fig 3.4

£m

VALUES AT CURRENT PRICES OF NEW ORDERS OBTAINED BY CONTRACTORS: BY REGION

North-West England

Scotland

South-West England

YEAR

3.2.1 Silhouette Charts

This is a line graph which shows how a variable has behaved relative to some standard amount (often zero). For example, we could

examine the fluctuations of someone's current account at a bank
relative to a balance of £0 with different shading for positive and
negative amounts.

Fig 3.5

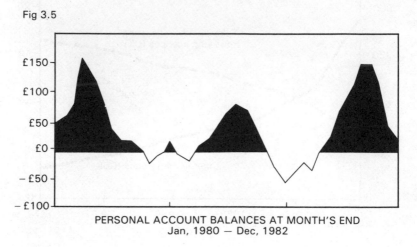

PERSONAL ACCOUNT BALANCES AT MONTH'S END
Jan, 1980 — Dec, 1982

3.2.2   Band Charts

This is another type of line graph where the variables recorded
are actually component parts of a total. Thus, each component
part will be shaded differently as in Figure 3.6.

In normal practice, the components are plotted in order of size
so that the largest component is on top. (Of course, a minor problem may occur if one component is largest for only some of the
period for which the data are collected. Any sensible resolution
of this minor problem is acceptable.)

3.3   Histograms and Bar Charts

In this section, we shall discuss histograms and bar charts. These
are probably the best known types of pictorial representation. Basically, a histogram or a bar chart is used when comparing the values
of several variables at a single point in time. However, variations
of histograms and bar charts do allow us to include values at a
few points in time. In these pictures we draw sets of rectangles
or blocks, one for each variable, and the heights of these blocks
are in proportion to the frequency/value of the variables.

In the following table, Table 3.1, we have data relating to the
Buyrite Corporation and their five divisions. We can plot this in

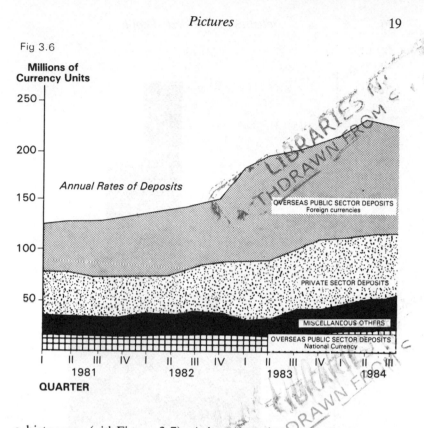

Fig 3.6

Millions of Currency Units

Annual Rates of Deposits

OVERSEAS PUBLIC SECTOR DEPOSITS
Foreign currencies

PRIVATE SECTOR DEPOSITS

MISCELLANEOUS OTHERS

OVERSEAS PUBLIC SECTOR DEPOSITS
National Currency

QUARTER

a histogram (vid Figure 3.7). A bar chart is essentially the same except that there may be a space between successive blocks.

Table 3.1

Buyrite Corporation

Division	Turnover in $m
A	7.7
B	11.4
C	24.6
D	19.3
E	9.0

We can now discuss some of the extensions and adaptations of bar charts. We can use bar charts to show breakdowns of finance, for example, as in Figure 3.8. Other uses are to show relative breakdowns with comparisons from year to year as in Figure 3.9.

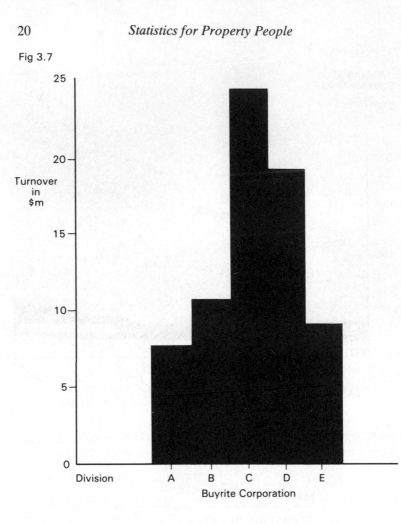

Fig 3.7

3.3.1   Component Bar Charts

In a component bar chart, we can look at the breakdown of some
total into its constituent parts as proportion of the total. Thus, in
Figure 3.10, we have a component bar chart showing the distribution of pupils in various types of schools. Nore that as each column
represents 100%, each column is of the same height. All the chart
shows is the change in pattern of distribution. For example, it is
possible that the number of pupils in grammar schools increases
in absolute terms but, because the total number of pupils is increasing faster, the percentage of pupils in grammar schools decreases.

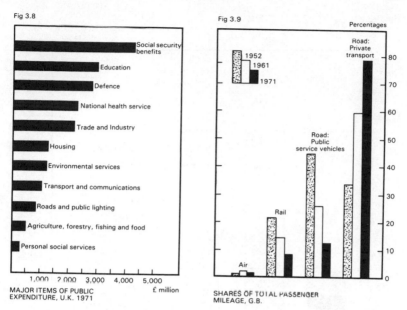

Fig 3.8

Social security benefits
Education
Defence
National health service
Trade and Industry
Housing
Environmental services
Transport and communications
Roads and public lighting
Agriculture, forestry, fishing and food
Personal social services

1,000 2,000 3,000 4,000 5,000

£ million

MAJOR ITEMS OF PUBLIC
EXPENDITURE, U.K. 1971

Fig 3.9

Percentages

1952
1961
1971

Road: Private transport
Road: Public service vehicles
Rail
Air

80
70
60
50
40
30
20
10
0

SHARES OF TOTAL PASSENGER
MILEAGE, G.B.

3.4 Pictograms

In a pictogram, we actually draw pictures of the items the data measure. For example, in Figure 3.11, we have a bar chart in pictorial form. This is from a set of employment statistics and this chart shows the increase in total man hours worked from 1880 to 1956.

Here, each full workman represents about 20 billion hours. Another example would be to plot data on population of English counties and use the outline of a man (or woman) to represent, say, 100,000 people.

3.5 Pie Charts

Another way to represent relative breakdowns is in a pie chart. In this, we draw a circle and partition it into segments, one for each component, with the size of each segment being proportional to size.

We can construct a pie chart for the data in Table 3.1. The circle is divided into 360° and the total turnover of the Buyrite Corporation is \$72m. The size of segment for Division A is $\frac{7.7}{72} \times 360° = 38.5°$. Similarly, for Division B, we have a segment

Fig 3.10

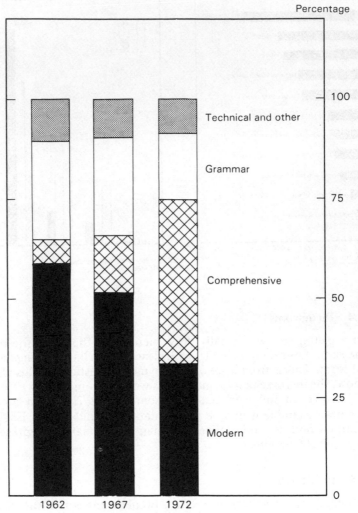

of $\frac{11.4}{72} \times 360° = 57°$. For Divisions C, D, E, we have segments of 123°, 96.5° and 45°, respectively.

The pie chart is in Figure 3.12 below. (Some standardization is desirable especially if one is to draw more than one pie chart for comparison. As a result, I suggest the following. Start at 12 o'clock

Fig 3.11

55 billion hours

148 billion hours

i.e. vertically, move clockwise and, unless there is some natural order, start with the largest segment, then the second largest, and so on.)

Fig 3.12

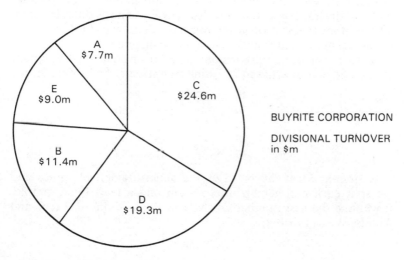

BUYRITE CORPORATION

DIVISIONAL TURNOVER in $m

3.6 Other Types of Pictures

There are many other types of pictures which may be of use. To display the classification into which different parts of a country fall, we may use a map with differential shading. Thus, to denote the political allegiances of the UK, we may plot the UK constituen-

cies on a map and use a different type of shading for those consti-
tuencies represented by Conservatives, Liberals, Labour members,
Independents, etc.

We can also develop other types of pictures especially by using
several of the ideas already mentioned in this chapter. For example,
differential shading or, even better, colour, will generally aid definition. Band charts, pie charts, and component bar charts will usually
benefit from good definition.

There is currently a new approach to Statistics which has been
growing greatly in support in the last ten years. This approach
is called Exploratory Data Analysis or EDA. Its basic idea is that
we examine and explore data to reveal things instead of trying
to impose our pre-conceptions. In other words, we try to use the
data to tell us what is happening rather than examining if the data
support our ideas about what should be happening. Part of EDA
is some rather new ways of looking at data. One of these new
ways is in a stem-and-leaf display.

3.7 Stem-and-leaf Displays

A recent development in displaying data is in the form of a stem-
and-leaf diagram or a stem-and-leaf display. This is different from
all the other types of diagrams we have dealt with in that we use
the actual numerical values as part of the diagram.

As an example, we can consider the following data on the number
of residential properties in 25 Edinburgh streets. These data are:

24	38	22	75	41
64	36	39	61	50
24	43	76	62	56
63	36	47	26	39
28	49	38	51	33

A stem-and-leaf display uses the actual numerical values and
we split each number up into a stem and a leaf. For example,
if we take the first number 24, we can think of this as 2 tens and
4 units; We can write this as

$$2 \mid 4$$

The next number is 38. We write this underneath the previous
number as 3 tens comes after 2 tens. Thus, we now have

$$2 \mid 4$$
$$3 \mid 8$$

The next number is 22. This must appear in the stem for 2 tens.

Thus, we place the 2 units after the 4 units from the 24. We can separate them by either a comma or by a space. Thus, we now have

$$\begin{array}{c|cc} 2 & 4 & 2 \\ 3 & 8 \end{array}$$

Continuing this until we have all 25 numbers, we get

$$\begin{array}{c|cccccccc} 2 & 4 & 2 & 4 & 6 & 8 \\ 3 & 8 & 6 & 9 & 6 & 9 & 8 & 3 \\ 4 & 1 & 3 & 7 & 9 \\ 5 & 6 & 0 & 1 \\ 6 & 4 & 1 & 2 & 3 \\ 7 & 5 & 6 \end{array}$$

To make this even clearer, it is necessary to add a title and a legend. Ordering the numbers in each stem may also help. Thus, we now have

Number of residential properties in 25 Edinburgh streets

$$\begin{array}{c|cccccc} 2 & 2 & 4 & 4 & 6 & 8 \\ 3 & 3 & 6 & 6 & 8 & 8 & 9 & 9 \\ 4 & 1 & 3 & 7 & 9 \\ 5 & 0 & 1 & 6 \\ 6 & 1 & 2 & 3 & 4 \\ 7 & 5 & 6 \end{array}$$

Stem = 10 residences
Leaf = 1 residence
e.g. 5 | 6 = 56 residences

We may now look at the advantages of a stem-and-leaf display. Firstly, a stem-and-leaf display is far quicker to construct than a histogram. This is because the data are being partially ordered automatically as opposed to a histogram where we really need the data to be ordered before we can begin. Secondly, it is more informative than a histogram as it retains more of the data and, therefore, contains more information. Thirdly, as a picture, it is just as useful as a histogram. In fact, turning the page on its side gives a shape which is identical to the histogram which would result from grouping the data in tens.

As was stated earlier, it is vital that a diagram be useful and clear. To amplify on this point, let us consider another data set. We have data on the rateable values of 30 properties in a town just outside Glasgow. The data are:

489	622	711	683	526	583
661	493	700	619	475	621
552	473	567	607	490	511
735	558	643	609	664	704
619	538	717	548	556	548

Several stem-and-leaf displays can be drawn to represent these data. For example, if we use hundreds of pounds as the stem and tens of pounds as the leaf then we get the display below. (Note that we may round up the figures to the nearest ten or merely ignore the units—the picture is essentially the same.) From this point, we shall separate the stem from the leaves, not by a vertical line, but by an appropriate symbol.

```
4  *  7  7  8  8  9
5  *  1  2  3  4  4  5  5  5  6  8          Stem = £100
6  *  0  0  1  1  2  2  4  6  6  8           Leaf = £10
7  *  0  0  1  1  3                      e.g. 5*4 = £540
```

We could draw another stem-and-leaf display by grouping the data in fifty's.

```
4  *  7  7  8  8  9
5     1  2  3  4  4
5  *  5  5  5  6  8
6     0  0  1  1  2  2  4
6  *  6  6  8
7     0  0  1  1  3            *now denotes "add 50"
```

The data can even be grouped in twenty's.

```
4  s  7  7
4  *  8  8  9
5  o  1
5  t  2  3
5  f  4  4  5  5  5          Here,
5  s  6                     o  denotes zero or one
5  *  8                     t  denotes two or three
6  o  0  0  1  1            f  denotes four or five
6  t  2  2                  s  denotes six or seven
6  f  4                     *  denotes eight or nine
6  s  6  6
6  *  8
7  o  0  0  1  1
7  t  3
```

Which one is used will depend on the context and on the type of information you may wish to be able to extract from the display. Another possibility is to put two digits in the leaf i.e. not to round the data. Thus, we could get the following display.

4	*	73	75	89	90	93			
5		11	12	38	48	48			
5	*	52	56	58	67	83			
6		07	09	19	19	21	22	43	Steam = £100
6	*	61	64	83					Leaf = £1
7		00	04	11	17	35			e.g. 5*56 = £556

Note that it is vital that we write the leaf for 704 as 04 and not as 4. The latter would lead to a distortion of the overall picture.

3.8 Ogives

In the previous section, we may not have been given all the information. For example, we may have been presented with the data on the number of residences in Edinburgh streets in the form of a frequency table e.g.

Number of Residences	Number of Streets
20–29	5
30–39	7
40–49	4
50–59	3
60–69	4
70–79	2

(Frequency tables will be dealt with in Chapter 5 but, briefly, this table states that 5 streets had between 20 and 29 residences in them, 7 streets had between 30 and 39 residences in them, and so on.)

With only this information, a stem-and-leaf is impossible to construct. A histogram would be the best alternative. However, another type of picture which would be useful is an ogive or a cumulative frequency graph.

Basically, we must first construct the cumulative frequency distribution which will be presented in a cumulative frequency table. None of these streets had less than or equal to 19 residences. Five of them had 29 or fewer residences. Twelve of these streets had 39 or fewer residences. Sixteen of them had 49 or fewer residences, and so on. Thus, we can now produce a table of the cumulative distribution.

Number of Residences	Cumulative Number of Streets
19	0
29	5
39	12
49	16
59	19
69	23
79	25

An ogive is merely a graph of these values with Number of Residences on the horizontal axis and Cumulative Number of Streets on the vertical axis. To make it clearer, the cumulative number is often replaced by percentage cumulative frequency which, in this case, as there are 25 streets, would be the values 0%, 20%, 48%, 64%, 76%, 92% and 100%. The points on the graph should be joined up by straight lines although any interpolation is hazardous. Further, for less than 19 residences, the corresponding percentage cumulative frequency is 0% while for more than 79 residences, this frequency is 100%.

Fig 3.13

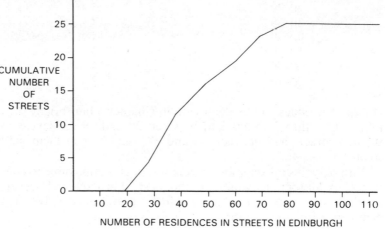

NUMBER OF RESIDENCES IN STREETS IN EDINBURGH

3.9 Good and Bad Practice

Most people think that Disraeli said "Lies, damned lies and Statistics". Actually, he did not. All that is recorded is that someone else reported him as saying it. However, if someone says that Disraeli said it, it is convenient to believe it. In a similar way, if a graph or a picture seems to show what it is convenient for you to believe or what you want to believe, you will do so. If the graph is persuasive by its presentation or misleading by its omission of facts which bias its presentation, you are even more likely to believe what the graph is trying to show.

Let us briefly consider some examples. In Figures 3.14a and b, there appear two graphs—of exactly the same data. The data are the total salary bill of a section of the US Government over a period of 8 years. In the first graph, the scale is very large so that one inch on the vertical axis represents $2.75 m. In the

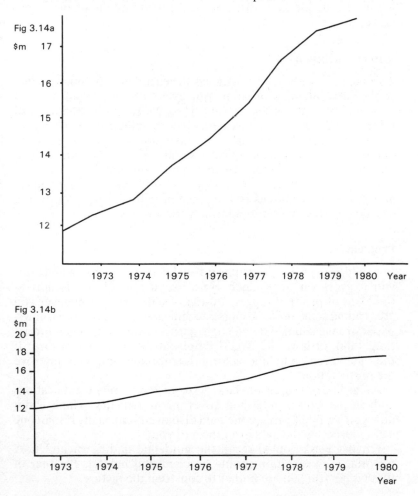

second graph, the scale is very small so that one inch represents $10 m. Which graph you choose may depend on what you are trying to show, rapid escalation or stability.

If your weekly pay rises from £80 to £160, has it risen by 100% or 50%? Obviously, it depends on whether you quote the rise as a percentage of the original figure or of the final figure.

There are many other situations in which ambiguities or misdirections can occur. I shall not discuss these but direct the interested reader to a delightful book which addresses this serious matter but in a light-hearted manner. The book is by D. Huff and is entitled "How To Lie With Statistics", published by Penguin. The text is full of examples and strategies for misleading the unaware and naïve.

3.10 Conclusion

This chapter is not really designed to teach you anything specific. Rather it is intended to open your eyes to the possibilities, both good and bad, of presenting data. H G Wells said that "Statistical thinking will one day be as necessary for efficient citizenship as the ability to read and write". We have not yet reached that stage and so, whatever statistical analysis you carry out or data you collect, there will come a time when you must present the information and/or your results to some non-specialist. Proper use of graphical and pictorial techniques is your responsibility. I urge you to use the ideas in this chapter to develop your own techniques.

Problems

It would not prove particularly fruitful to provide examples for you to work on. It is much more useful for you to be alert to the ideas of pictorial representation of data. Thus, whenever you are reading some professional or technical material or even a newspaper or magazine, and you come across some numerical information, think of how you would represent the data in a picture. If the data are already in a pictorial form, think of how to improve the presentation.

As an extra source of data, look at government publications and, in particular, published government statistics. Think about how you would like to see the data displayed; especially if someone were presenting the data in a report to you.

Finally, I ask you to recall the guideline laid down in 3.1. A pictorial representation should be clear and helpful. If either of these is not true, do not bother to construct the picture.

PART THREE
The Basic Language

CHAPTER 4

Language and Conventions

4.1 Introduction

Any field of study has its own jargon and "buzz-words" and language and conventions. As these terms are absorbed into everyday English, they hold less and less fear. For example, most people now understand that an analgesic is a pain-killer and that there is nothing wrong with a pressurized cabin in an aeroplane.

Statistics also has its technical terms and language and conventions. Some of them are mathematical and some are conventions of English but they all must be understood if a thorough knowledge of Statistics is to be gained. These terms, symbols and abbreviations are difficult at first but there are clues to their use and meaning. This chapter, therefore, is designed to introduce much of this jargon before we really need to use it. As a result, this chapter consists of disjoint fragments. Do not, however, let that dishearten you.

4.2 Symbols

As was stated earlier, statistics deals with population parameters and the knowledge we gain from sample statistics. We generally use Greek letters to denote population parameters and Roman letters to denote sample statistics. The most common are listed below. Their meaning will not be clear but it is important that, at an early stage, we grasp the correspondence between a Greek letter and a Roman letter. The population parameters are read as "mu", "sigma squared", "sigma", "pi", and "rho".

Population		Sample
μ	MEAN	$\bar{x}$
σ^2	VARIANCE	s^2
σ	STANDARD DEVIATION	s
π	PROPORTION	p
ρ	CORRELATION	r

4.3 Verbal Quantifiers

In this section, we shall be considering what I call verbal quantifiers. These are such terms as

"and"   "or"
"at least"   "less than"
"at most"   "more than"

These are all clearly defined in English but as soon as they are put in a mathematical type question, most students have difficulty in understanding them.

We shall consider each of them in turn and in the context of the following example: I shall go to France for my holidays if . . .

and: if the Socialist Government is re-elected and the Franc is devalued.

In this case, I shall go to France only if both these things occur. If the Socialist Government is re-elected but there is no devaluation I shall not go. Neither shall I go if the Socialist Government is not re-elected and the Franc is devalued. Only if both things occur is the "and" satisfied.

or: if I can get a cheap plane fare or the Franc is devalued.

In general, we shall use what in logic is called the inclusive "or". Therefore, I shall go if I get a cheap plane fare and there is no devaluation. I shall also go if I cannot get a cheap plane fare but there is a devaluation. I shall also go if both occur.

at least: if at least three people recommend it to me.

Here, I shall go if six people recommend it to me or five people recommend it to me or four people recommend it to me or three people. Three is, after all, at least three.

Usually we can re-express "at least" as "or more". Thus, "at least three" is equivalent to "three or more".

more than: if more than three people recommend it to me.

Here, I shall go if four or five or six (or any larger number of) people recommend it to me. However, if only three recommend it, I shall not go as three is not more than three.

at most: if I hear of at most four people getting upset stomachs.

Here, I shall go if zero or one or two or three or four people get upset stomachs.

less than: if I hear of less than four people getting upset stomachs.

Thus, I shall go if I do not hear of anyone or if I hear of one or two or three people. However, if I hear of four people, I shall not go as four is not less than four.

These are the most common verbal quantifiers. Be careful with them.

4.4 Logarithms

10^2 equals 100. Therefore, the logarithm of 100 to base 10 is 2. We write this as

$$\log_{10}(100) = 2$$

$10^{0.74}$ equals 5.4954. Therefore, the log of 5.4954 to base 10 is 0.74.

$$\log_{10}(5.4954) = 0.74$$

Usually, if we are referring to logarithms to base 10, we write log without a 10 as a subscript. Most calculators also use this. However, there is one other base which is commonly used for logarithms. This is for natural logarithms and is base e where e is $2.7\dot{1}8\dot{2}\dot{8}$. (The dots above the numbers indicate that this sequence of four digits is repeated ad infinitum.) We generally denote natural logarithms by ln

$$\ln(100) = 4.60517 \text{ i.e. } (2.7\dot{1}8\dot{2}\dot{8})^{4.60517} = 100$$
$$\ln(5.4954) = 1.7039$$

In fact, we can convert a logarithm to base 10 to a natural logarithm by dividing by log (e) which is approximately 0.43429.

4.5 The Equation of a Straight Line

In Mathematics, Statistics, Economics and several other subjects, much use is made of straight lines and straight line relationships between variables. Therefore, this text uses straight lines and we need to clarify the notation and formula.

If we have a regular Cartesian co-ordinate system, we may have two points on it. One is at $(2, 3)$ and the other is at $(4, 5)$. If we draw a straight line through these two points, what is the equation of this line?

We specify lines by a general form of

$$Y = mX + C$$

Where X and Y are the first and second co-ordinates, i.e. horizontal and vertical co-ordinates of all points on this line and m is the gradient or slope of the line and c is what is called the y-intercept or the vertical co-ordinate where the line cuts the vertical axis.

Fig 4.1

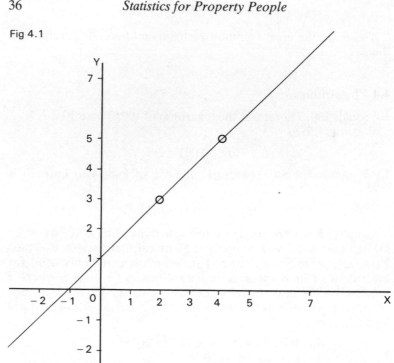

How do we evaluate m and c? m is the gradient and we generally have the gradient $= \dfrac{\text{difference in } Y \text{ co-ordinates}}{\text{difference in } X \text{ co-ordinates}}$

Here, $m = \dfrac{5-3}{4-2} = \dfrac{2}{2} = 1$

For the y-intercept, we take either of the points and solve for $Y = mX + c$. In this case, we have $3 = 1 \times 2 + c$ and so $c = 1$.

Thus, the line has equation $Y = X + 1$ and all points on this line or extensions of this line will satisfy this equation.

If we had the points $(2,2)$ and $(1,4)$, the line through these two points has gradient $\dfrac{4-2}{1-2} = -2$ and the intercept is 6. Thus, the equation of the line is $Y = -2x + 6$ or $Y = 6 - 2X$.

4.6   Combinatorics

4.6.1   Factorials

We now introduce the factorial sign ! which, as you can see, is identical to an exclamation mark. We use this with non-negative integers. Thus, we can write 4! or 6!. We read this as "four factorial" and "six factorial".

The values of these factorials are as follows

For 4!, we evaluate this as $4 \times 3 \times 2 \times 1 = 24$
For 6!, we evaluate this as $6 \times 5 \times 4 \times 3 \times 2 \times 1 = 720$
For 7!, we evaluate this as $7 \times 6 \times 5 \times 4 \times 3 \times 2 \times 1 = 5040$

$$2! = 2, \quad 1! = 1, \quad 3! = 6.$$

Note that 0! is, by convention, equal to 1.

4.6.2 Combinations

Let us suppose that we have five different objects, and we wish to select two of them. In how many ways can this be done? If we label the objects $A - E$, there are ten ways of selecting two:

viz			
	AB	BC	CD
	AC	BD	CE
	AD	BE	DE
	AE		

Such a method is messy and wasteful of space and, for large numbers, very time consuming. However, we can use the factorials above in a compact formula for the number of possible selections.

We have 5 objects and wish to select 2 and discard (or not select) 3. The number of possible selections is

$$\frac{5!}{2!\,3!} \quad \text{or} \quad \frac{5!}{2!(5-2)!} = \frac{120}{2 \times 6} = 10$$

Suppose that we wished to select 3 objects out of 8. The number of selections is

$$\frac{8!}{3!(8-3)!}$$

We can simplify the calculation by noting that $8! = 8 \times 7 \times 6 \times 5!$

$$= \frac{8 \times 7 \times 6 \times 5!}{3! \times 5!} = \frac{8 \times 7 \times 6}{3 \times 2 \times 1} = 56$$

We have a notational form for this. It is $\binom{8}{3}$ or, in the previous case $\binom{5}{2}$. These numbers also appear in Pascal's Triangle. Each value in the triangle is the sum of the two values above it. Where there is no value, a zero is assumed. If we look at the fifth row, it is 1 5 10 10 5 1. These correspond to $\binom{5}{0}$, $\binom{5}{1}$, $\binom{5}{2}$, $\binom{5}{3}$, $\binom{5}{4}$, $\binom{5}{5}$ where by $\binom{5}{0}$ we understand the number of ways of selecting none of five items which is to throw them all away i.e. there is only one possible selection.

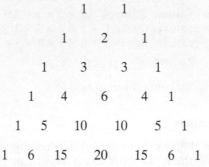

$$
\begin{array}{ccccccc}
 & & & 1 & & 1 & & & \\
 & & 1 & & 2 & & 1 & & \\
 & 1 & & 3 & & 3 & & 1 & \\
 1 & & 4 & & 6 & & 4 & & 1 \\
1 & 5 & & 10 & & 10 & & 5 & & 1 \\
1 & & 6 & & 15 & & 20 & & 15 & & 6 & & 1
\end{array}
$$

In these selections, we have implicitly assumed that the order is unimportant i.e. a selection of object A then object B is the same as selecting object B then object A. This gives rise to the $\binom{n}{x}$ formula for selecting x objects from n (different) objects. This is called the number of combinations.

[Note that $\binom{n}{x} = \binom{n}{n-x}$. Can you prove this?]

4.6.3 *Permutations*

If the order of selection is important so that AB and BA constitute two different selections, then the total number of possible selections is the number of permutations. (Note that when the football pools refer to permutations, they are referring to combinations.)

The formula for the number of permutations of x objects from n different objects is

$$
\frac{n!}{(n-x)!}
$$

Thus the number of permutations of 2 objects from 5 is 20. (The ten listed above plus ten with the order reversed.) The number of permutations of 3 objects from 8 is $\frac{8!}{5!} = 336$.

4.7 Subscripts

In most scientific subjects, letters are often used to denote measures. For example, we may be considering a domestic property.

We will record several measures such as floor area, number of rooms, rateable value, etc. We could call the floor area A, the number of rooms B, and the rateable value C. If we were to consider a number of properties, five, say, we would have five values of each of A, B, and C.

To distinguish the measures of floor area for the five properties, we often use subscripts—numbers or letters written below the writing line. Thus, if we number the properties 1 to 5, we can refer to the floor areas as A_1, A_2, A_3, A_4, and A_5. (This can be read as "A sub 1", "A sub 2", etc. or just as "A 1", "A 2", etc.) Similarly, the rateable values would be denoted C_1, C_2, C_3, C_4, and C_5.

Often, particularly if there is a large number of observations (in this example, a large number of properties), we use a letter to denote each of the numbers. Thus, instead of writing B_1, B_2, B_3, B_4, and B_5, we may write B_i, $i = 1, 5$. (This can be read as "B sub i", where i takes values from 1 to 5.) Sometimes, we may drop the "$i = 1, 5$" in contexts where it is clear what values i takes.

4.8 Summation

We can use the subscripts developed above, to produce a shorthand notation which will be used frequently when we wish to add up a set of numbers. This notation involves the Greek letter sigma again, but this time it is in its upper case or capital form Σ.

In mathematics, Σ often denotes a sum or a summation. Thus, if we again consider the floor areas of the five properties in §4.7, we may see how this notation can be used to denote the sum of the floor areas.

The sum of the floor areas could be written as $(A_1 + A_2 + A_3 + A_4 + A_5)$. However, a neater notation, and particularly so if there are a larger number of properties being considered, is

$$\sum_{i=1}^{5} A_i.$$

The Σ can be read as the sum and the whole expression can be understood to read "As i takes values from 1 to 5, take the values of A_i and add them up".

Often, however, we shall be dealing with x's and y's instead of A's, etc. and so, in looking at an example of this notation, let us take four values of x_i:

$$x_1 = 2, \qquad x_2 = -5, \qquad x_3 = 1, \qquad x_4 = 3.$$

$\displaystyle\sum_{i=1}^{4} x_i$, is the sum of x_1, x_2, x_3, and x_4, and equals 1.

$\displaystyle\sum_{i=3}^{4} x_i$, is the sum of x_3 and x_4 only, and equals 4.

$\displaystyle\sum_{i=1}^{4} x_i^2$ is the sum of the squares of x_1, x_2, x_3, and x_4, and is equal to 39.

$\displaystyle\sum_{i=1}^{3} x_i^3$ equals -116. (Check this for yourself.)

Finally, we must consider $\displaystyle\sum_{i=1}^{4} 2$. It may seem rather odd but, when $i = 1$, the number to be added is 2. When $i = 2$, the number to be added is 2, and so on. Thus, we are adding up 2 four times and the answer is 8.

4.9   Absolute Values

Sometimes, it may be necessary to take the absolute value of a number. Simply speaking, the absolute value of a number is its size. Thus, the absolute value of a positive number is itself. The absolute value of zero is zero. The absolute value of a negative number is the positive version of this number i.e. the number multiplied by -1. This is because a size cannot be a negative number.

The absolute value of a number, x, is usually denoted by an x with a vertical bar on either size e.g. $|x|$.

Thus, $|3| = 3$
$\qquad |28.3| = 28.3$
$\qquad |0| = 0$
$\qquad |-16| = 16$
$\qquad |-5.8| = 5.8$

Note that the absolute value is sometimes referred to as the modulus.

Problems

1. What is the value of $\binom{17}{3}$, $\binom{8}{1}$, $\binom{6}{0}$, $\binom{21}{2}$, $\binom{40}{3}$, $\binom{4}{4}$,?

2. An estate agent had 25 flats for sale at the beginning of September and managed to sell 6 of them by the middle of October. How many ways are there to select 6 flats from the 25?

3. A used car salesman has 14 cars in his yard. To stimulate business, he offers £50 off the price of the first of these cars to be sold, £40 off the price of the second one, and so on until £10 off the price of the fifth one. In how many ways can 5 cars be selected from the 14 to receive these special offers?

4. Four lawyers are to be selected from a group of 38 to represent the group in a meeting of city professionals. How many ways are there to choose the four representatives?

5. If $y_1 = 3$, $y_2 = 11$, $y_3 = -2$, $y_4 = 5$, and $y_5 = 1$, evaluate the following sums.

$$\sum_{i=1}^{5} y_i, \quad \sum_{i=1}^{5} y_i^2, \quad \sum_{i=1}^{5} y_i^3$$

$$\sum_{i=1}^{5} (y_i - 2)^2, \quad \sum_{i=1}^{3} y_i^2, \quad \sum_{i=1}^{3} |y_i - 6|,$$

$$\sum_{i=3}^{5} (y_i - 2)^3.$$

CHAPTER 5
Descriptive Statistics

5.1 Introduction

This chapter will deal with descriptive or summary statistics. These statistics do just what their name implies; they describe and summarize a set of data. They do not, of themselves, present analysis or offer conclusive answers although they may be used in an analysis to arrive at conclusive answers. Rather, they offer a method of identifying and reporting the salient features of a data set.

We shall develop descriptive statistics to look at three properties of a data set. These are the centre, the spread, and the shape.

The data set which we are going to consider is part of a survey carried out in 1981 of some people who were just about to move out of privately rented property. They were asked "How long (to the nearest month) have you been living in your rented accommodation?"

The responses from 17 people were

$$26 \quad 9 \quad 23 \quad 6 \quad 23 \quad 10 \quad 74 \quad 15 \quad 13 \quad 14$$
$$12 \quad 17 \quad 6 \quad 5 \quad 27 \quad 6 \quad 20$$

If we wished to report the results of this survey, quoting the seventeen figures would be untidy and wasteful—even more so if the data set were far larger. Graphs and pictures are one way to convey information about the data but summary statistics also provide some of the first steps towards analysis.

5.2 The Centre

We must be able to describe the centre of the data set or, in more general terms, the location of the numbers. We must be able to say whether the numbers are situated around -64 or $2,001$ or 42. We do this by replacing the data by a representative number; an "average". I have highlighted the word average as there are many different types of average. The various types convey different types of information and have their own advantages and disadvantages. We shall deal principally with three types of averages—the mean, the median, and the mode.

5.2.1 The Mean

The mean or, to give its proper name, the arithmetic mean is, in fact, what most people understand by the average. It equals the sum of all the values divided by the number of values. In symbolic terms, it can be written as

$$\frac{\sum_{i=1}^{n} x_i}{n}$$

Thus, in the data above on the time in rented property, the mean is the sum of the values (306) divided by the number of values (17). This give a mean time of 18 months.

The mean has advantages and disadvantages. The main advantage is that most people have some intuitive feel for it. The main disadvantages are

(i) it can be difficult to calculate, especially for large data sets.

(ii) the mean may not be one of the values which occurred; no one actually responded with 18 months. Further, it may not even be a value which could have occurred; the mean number of children per family in an area is 2.33.

(iii) the mean may be adversely affected by one or two extreme values; if the response of 74 had been omitted, the mean would be 14.5 months.

(iv) the mean may be inappropriate in certain situations.

5.2.2 The Median

Simply speaking, the median is the middle number i.e. there are as many numbers smaller than the median as there are larger than it. Thus, to get the median we need to have the data in order or at least have the middle section in order. (Alternatively, looking at a stem-and-leaf display may be of help here.)

For example, suppose the values in a data set were 16, 22, 9, 17, and 20. The median is 17 as there are two numbers smaller than it and two numbers larger than it. In fact, any time that we have an odd number of values in a data set, the median is easily identified as the middle value. However, if there are an even number of values in a data set, then we take the two values in the middle and compute their mean. For example, if the above data set of five values had a value of 19 added, then, in order, the values would be 9, 16, 17, 19, 20, and 22. The median can be thought of as lying between 17 and 19. (One can even think of drawing a line between 17 and 19 to divide the data set into two groups each of size three.) In theoretical terms, any number between 17 and 19 is valid as the median of this data set but we shall follow the generally accepted rule of taking the mean of the two values in the middle and get a median of 18.

We can develop a rule of thumb for finding the median and later on we can use this rule to find other statistics. Suppose that we have n values in a data set. The median is the half-way point and so we divide n by 2 to get $n/2$.

If $n/2$ is not a whole number, round it up to the next whole number.

If $n/2$ is a whole number, add $\frac{1}{2}$.

Thus, if we have data set with 5 values, we take $\frac{5}{2}$ and then round it up to the next whole number ($2\frac{1}{2}$ becomes 3) and so the median is the third number along the ordered list (third smallest or third largest gives the same value). On the other hand, when we had a data set of size 6, we take $\frac{6}{2}$ which is 3 and then add $\frac{1}{2}$ to get $3\frac{1}{2}$. Whenever we see a $\frac{1}{2}$ in this situation, we understand it to mean that we take the two values on either side and take their mean. Thus, for $3\frac{1}{2}$, we take the third and fourth values on the ordered list and compute their mean. (Once again, it does not matter which end of the ordered list we begin to count from.)

Check that, in the example on rented property, you get a median of 14 months.

The median also has disadvantages. One is a purely technical point in that we cannot really begin to calculate the median until we have all the observations available. A second problem is that, like the mean, we may get a median which either did not occur or could not have occurred. A third problem is that, as yet, people do not have as great an intuitive feel for the median as they have for the mean. The most important advantage that the median has is that it is not affected by one or two extreme values. Thus, the median would not be altered if the value of 74 were changed to 24 or to 124.

5.2.3 The Mode

Basically, the mode is the value which occurs most often. In the survey results, the value 6 occurred three times and the value of 23 occurred twice. All the other numbers occurred once. Thus, the mode is 6 months.

The mode does, obviously, have to be one of the numbers which occurred and is easy to calculate—it is merely a matter of identification—but it does have some disadvantages.

The main disadvantage is that there may be more than one mode in a data set. For example, if the value of 23 had occurred three times, there would have been two modes—at 6 months and at 23 months. This can be a source of confusion but, in some situations, this can be informative. For example, it may be that we have been able to identify two different types of people here; one with a mode of 6 months and another with a mode of 23 months.

Another disadvantage is that, if the data are presented in the form of a frequency table, we are unable to identify or even estimate the mode. (We can do this for the mean and median and shall do so below). We can easily identify the modal group i.e. the most popular group but there is no guarantee that the mode falls within the range of the modal group.

5.2.4   Mean, Median, and Mode—which one?

Before we look at some examples where one of the three might be deemed inapplicable, two points ought to be made. Firstly, the three summary statistics are only the three most common among hundreds of summary statistics, many of which have been developed mathematically in an effort to avoid some of the disadvantges stated above. Secondly, these three summary statistics convey different types of information and, therefore, it is not necessarily redundant to quote more than one of them. Indeed, comparison may prove informative.

If you were to ask a retailer "What is the average size of dress you sell?", it may seem that the mode is more informative than the other two. If you were a supplier, then supplying the modal size of dress allows the retailer to satisfy demand as much as possible. The median may be reliable depending upon the population and their eating habits—unreliable in a region where women tend to be either very thin or overweight—and the mean size does not make sense as dress sizes 8, 10, 12, 14, etc. are merely codes.

Another example might be to ask "What is the average man's income?" Here, the mean will be distorted by a few very large incomes (the 7:84 rule). The mode from a sample may just be a fluke of two men earning the same amount while the median is probably the most informative and reliable.

Returning to the example on dress sizes, if a dress manufacturer asks "What is the average amount of cloth in a dress?", it might be best to quote the mean. This gives the information the manufacturer will need if, for example, he has to decide how much cloth to buy to fill an order for, say, 8,000 dresses.

If an estate agent wished to know the average house size, then we have to decide which type of average to quote. The mode will give the estate agent information about the most frequently occurring size of house. The mean may be of importance if he wishes to know the average physical size of a house so as to plan an entire housing estate. The median will, of course, tell him about the middle of the range. Which one, if any, is going to be of use to him? It really all depends on what the purpose is; that is the key to choosing an appropriate average.

5.2.5   Mean, Median, and Mode for Grouped Data

Two hundred married couples were asked "How many rooms, excluding kitchen and bathrooms, do you have in your home?" The results were as follows.

Table 5.1

Number of Rooms	Number of Couples
1	4
2	15
3	34
4	42
5	67
6	28
7	6
8	3
9	0
10	0
11	1
	200

From this table, it is easy to identify the mode of 5 rooms. For the median, we need to construct a third column with cumulative frequencies (as in Figure 5.2). From this third column, it is easy to see that both the 100th and 101st values are 5. Thus, the median is 5 rooms.

The mean, however, requires a slightly different approach. We need to get the sum of the two hundred responses to the question. Consider the couples with seven rooms. There are six couples with seven rooms. We could add up seven six times i.e. $7 + 7 + 7 + 7 + 7 + 7 = 42$. Rather, I hope that we would simply multiply 7 by 6. (After all, multiplication is merely repeated addition.) This process must be carried out for each of the responses. Thus, we can add a fourth column to the table (as in Table 5.2).

Table 5.2

Number of Rooms x_j	Number of Couples f_j	Cumulative Frequency	$x_j f_j$	$x_j^2 f_j$
1	4	4	4	4
2	15	19	30	60
3	34	53	102	306
4	42	95	168	672
5	67	162	335	1,675
6	28	190	168	1,008
7	6	196	42	294
8	3	199	24	192
9	0	199	0	0
10	0	199	0	0
11	1	200	11	121
			884	

We have now calculated the sum as 884 rooms and this gives a mean of 4.42 rooms.

If we use the subscript j to denote the k classes in the frequency table, we can write the mean in symbolic terms as

$$\frac{\sum_{j=1}^{k} x_j f_j}{n}$$

Now, if we consider a different data set on an estate agent's last twenty sales, we run into a different problem.

Table 5.3

Price in £000's (to nearest thousand)	Number of Sales
0–9	1
10–19	3
20–29	7
30–39	5
40–49	4
	20

Now, as the data are grouped, we have lost the exact information on the sales. As a result, we cannot estimate the mode with any accuracy. However, we can conclude that the modal group is the group which goes from twenty to twenty-nine thousand pounds. (It must be realized that the modal group may depend on exactly how we split up the data into groups.) The median occurs in the same group and could be estimated fairly accurately using an ogive and reading off the value on the ogive against 50% on the vertical scale.

Once again, the mean causes some problems in calculation. There are four values in the 40–49 group but we do not know exactly what they are. We cannot accurately guess what they are but we can use some number to represent them. It is usual to use the mid-point of the group. Thus, we can take 44.5 as the midpoint (or the mean of the end-points). The same applies for the other groups (34.5, 24.5, 14.5) but there is a practical problem with the 0–9 group. In practical terms, the smallest number which can contribute to this group is a price of £0 while the maximum price which will usually be counted in this group is £9,500. Thus, strictly speaking, the mid-point is £4,750. (This slight adjustment can often be ignored provided one proceeds with caution.) If we use these estimates, we can rewrite the data as in Table 5.4. We now have the data in a form which we can cope with as we did above with the data on the number of rooms in houses. The estimate of the mean should be a price of £28,512.50.

Table 5.4

Price (estimate in £000's)	Number of Sales	Cumulative Frequency	$x_j f_j$	$x_j^2 f_j$
4.75	1	1	4.75	22.5625
14.5	3	4	43.5	630.75
24.5	7	11	171.5	4,201.75
34.5	5	16	172.5	5,951.25
44.5	4	20	178	7,921

5.3 The Spread

Just as we must be able to describe the centre of the data, we must also be able to describe the spread of the data. For example, if we have a mean or median of 23, it may be that all the values lie between 21 and 25 or between 12 and 31 or between −14 and 86. We need to have some method of measuring how widely spread the data are. We shall discuss five measures of spread which, as with measures of the centre, convey different types of information. The five measures to be discussed are the range, the mean absolute deviation, the variance, the standard deviation, and the inter-quartile range.

5.3.1   The Range

In the example on rented property, the easiest measure of spread is to say that the values go from 5 to 74. The range of these values is merely the difference between these two and we have a range of 69 months. More formally, the range can be defined as the difference between the largest (or maximum) value and the smallest (or minimum) value. It is easy to calculate but does not convey much information as it only involves two of the values. It does not really tell us how far from the centre we might expect to find a typical value. (Further, the range is easily influenced by an odd value.)

5.3.2   Deviations from the Centre

As we are really attempting to describe how far from the centre we might expect to find a typical value, we ought to be dealing with deviations from the centre, i.e. arithmetic differences between each value and the centre of the values.

We can take the values (which we shall label x_i) and subtract from each of them the mean (labelled $\bar{x}$) and call this the deviation (labelled $d_i\,(=x_i-\bar{x})$). Traditionally, calculations of this sort have used the mean as the centre. Only recently have people begun to use other measures of the centre. Until other measures are used by non-statisticians, it would seem wasteful to spend time discussing such approaches in depth.

One thing to do with these deviations is to find their mean. However, this mean will be zero—whatever the data. To prove it to yourself, take any five or six numbers and calculate the mean deviation of these numbers from their mean. If you do not get them to add up to zero, then you have made an arithmetic mistake. In fact, one way to define the mean is that it is the only number which has this property. To return to the problem, if we cannot find the mean of the deviations, what can we do?

5.3.3   The Mean Absolute Deviation

The mean of the deviations is zero because positive and negative deviations cancel each other out. Therefore, if we can remove these cancellations, we may be able to get a meaningful measure of deviation. One way to try is to look at the absolute value of the deviations. If we can calculate the mean of these absolute values, we call this the mean absolute deviation. Symbolically, the mean absolute deviation can be written as

$$\frac{\sum\limits_{i=1}^{n} |x_i - \bar{x}|}{n}$$

In this example, the mean absolute deviation is equal to

$$\frac{(8 + 9 + 5 + \ldots + 9 + 12 + 2)}{17}$$

which equals 10 months.

5.3.4 The Variance

Another way to get round the positive and negative deviations cancelling each other out is to square all the deviations before evaluating their mean. This would be the mean squared deviation and can be written symbolically as

$$\frac{\sum\limits_{i=1}^{n} (x_i - \bar{x})^2}{n}$$

However, for theoretical reasons which we shall not discuss here, we generally divide by one less than the number of deviations i.e. by $n - 1$. This calculation gives the variance and, as was mentioned earlier, we denote the variance by σ^2.

In this example, the variance is equal to

$$\frac{(8^2 + (-9)^2 + 5^2 + \ldots + 9^2 + (-12)^2 + 2^2)}{16}$$

which equals 260.5.

There are at least two things to be noted here. The first one is that the calculation was long and the second one is that this measure of spread is far larger than the mean absolute deviation evaluated above.

This method of calculation is, in fact, inefficient because we went through the seventeen values once to get their mean and then went through them again to get the deviations from the mean. There is another formula for the variance which is far quicker computationally. The original formula which we used above can be written in symbolic terms as

$$\frac{\sum_{i=1}^{n} (x_i - \bar{x})^2}{n - 1}$$

The alternative formula which is often known as the computational formula for the variance can be written in symbolic terms as below.

What this says is that we only need to go through the values once. However, we keep a running total of two things—the sum of the values and the sum of the squares of the values. We take the sum of the values, square it and divide by the number of values. We then subtract this from the sum of the squares and divide this difference by one less than the number of values.

$$\sigma^2 = \frac{\sum_{i=1}^{n} (x_i^2) - \dfrac{\left(\sum_{i=1}^{n} x_i\right)^2}{n}}{n - 1}$$

Now, one does not have to actually keep two totals running simultaneously and even working out the two sums separately seems a lot more complicated than the original formula. However, it is a lot faster. With our data, we have 17 values. The computational formula requires us to carry out only one subtraction while the original formula requires seventeen subtractions. If we had even more observations, the saving in time would be even greater. In our example, $\Sigma(x_i^2) = 9{,}676$ and $\Sigma x_i = 306$ and entering these into the formula gives a variance of $\dfrac{9{,}676 - (306)^2/17}{16}$ which equals 260.5.

This measure is far larger than the mean absolute deviation. That is because we are really working in different units. Each of the values and their mean are measured in months as are the deviations. However, once we square the deviations, the units are "squared months". We then take a type of average which does not change the units. Thus, the units of our variance are "squared months". In other situations, we may have to deal with a variance measured in "squared dollars" or "squared towns" or "squared people". The way to avoid this is to take the square root of the variance. This measure will now be measured in the same units as the original values. This measure is called the standard deviation.

5.3.5 The Standard Deviation

In computational terms, the standard deviation is the square root of the variance which, in this case, is equal to 16.14 months. While this is another measure of how far, on average, the data are from their mean, we can quote some rules of thumb which provide some idea of how to interpret the standard deviation.

(i) In most data sets, about 68% of the values lie within one standard deviation of the mean.
(ii) In most data sets, about 95% of the values lie within two standard deviations of the mean.
(iii) In all data sets, at least 75% of the values lie within two standard deviations of the mean.
   (How these rules of thumb arise is dealt with elsewhere in the text.)

5.3.6 Percentiles—Quartiles

A percentile is a value which is cumulatively so many per cent along the data set. For example, the 35th percentile (sometimes written as the 35%-ile) is the value which is greater than 35% of the values and is exceeded by 65% of the values.

We have already met one of the percentiles. That was the median; for the median is the 50th percentile. However, other percentiles are often of great importance. For example, certain percentiles appear in designing structures. We may be interested in fixing a shelf on a wall. How high do we fix it? There is no point in fixing the shelf at a height which the average person can barely reach. Whatever measure of average we use, there will be a large group of people who cannot reach the shelf. Clearly we wish to fix the shelf at a height which allows most people to reach it comfortably. Thus, we might choose a height which can be reached by 85% or 90% or 96% of people. Thus, if we look at the heights that people can reach, we would pick out the 85%-ile or the 90%-ile or the 96%-ile. This type of approach is used in fixing the heights of doors, in designing cars, and in several other areas where anthropometric measurements—measurements of human beings while moving around doing things—are involved.

As well as the median, at least two other percentiles are often used. These are the 25th percentile and the 75th percentile. The 25th percentile is a quarter of the way along the cumulative frequency and it divides the data into two groups—about one quarter of the values are less than it and about three quarters of the values are greater than it. For this reason, it is also known as a quartile and, in fact, the 25%-ile is the lower quartile. The 75th percentile also divides the data into two groups in the proportions $\frac{3}{4}:\frac{1}{4}$ and it is known as the upper quartile.

From time to time, other percentiles may be of interest and below we shall again use the rule stated above for finding the median. However, we can use it to find any percentile which is of interest. Note that a more general way to find a percentile is from an ogive.

5.3.7   The Inter-Quartile Range

Another measure of spread is the inter-quartile range. This is the difference between the upper and the lower quartiles i.e. it is the range of the middle half of the data set. It has an obvious advantage over the full range in that it is not affected by one or two extreme values. However, it is not obviously any better than using the range of the middle two-thirds of the data set or the middle five-eighths. In fact, we really use the middle half to be consistent with current statistical practice.

The rule we use to find a percentile is a generalization of the rule we used in 5.2.2 to find the median.

If n is the size of the data set, for the lower quartile, we wish to find the 25th percentile. Thus, we take a quarter of n.

If $n/4$ is not a whole number, round it up to the next whole number.

If $n/4$ is a whole number, add $\frac{1}{2}$.

(Once again, we understand $5\frac{1}{2}$ to be the mean of the fifth and sixth numbers on the ordered list.)

This process can also be carried out for the 75%-ile or for any other percentile.

In the data on rented property, n is 17. For the lower quartile, 25% of 17 is 4.25 which is rounded up to 5. We can see that, if we take the lower quartile to be the fifth smallest value, there are four values smaller than it and twelve values larger than it. Similarly, the upper quartile results from taking 75% of 17 to get 12.75 and then rounding this up to 13. (Note that the upper quartile could also have been found by looking at the fifth largest value.) Thus, the lower quartile is equal to 9 months and the upper quartile is equal to 23 months and the inter-quartile range is 14 months.

5.3.8   Measures of Spread for Grouped Data

If we consider the data which appear in Figure 5.1, we can try to calculate various measures of spread.

The range is the easiest to calculate as it is the difference between 11 and 1 and is, therefore, 10 months.

For the standard deviation, we can think of the data in the same way as we did when we discussed the mean for this data set. We have four responses of 1 room. Whatever the contribution each of these responses makes to the standard deviation, these contributions must be identical. Thus, we calculate their contribution once and multiply by the frequency of the identical responses which, in this case, is four. This procedure is repeated for all the responses and we can write this in symbolic terms as

$$\frac{\sqrt{\sum_{j=1}^{k} f_j(x_j - \bar{x})^2}}{n - 1}$$

Once again, there is a computational form which is far quicker to use when actually doing the calculation.

$$\sigma = \frac{\sqrt{\sum_{j=1}^{k} f_j x_j^2 - \left(\sum_{j=1}^{k} f_j x_j\right)^2 / n}}{n - 1}$$

To assist this calculation, we can use the last two columns of Table 5.2.

Thus, the standard deviation equals the square root of the variance which equals

$$\frac{4,332 - (884)^2/200}{199}$$

The variance equals 2.13427 and the standard deviation is 1.4609 rooms.

The mean absolute deviation can be calculated using the same principle—that equal values must make the same contribution to the answer. You should be able to calculate the mean absolute deviation to be 1.159 rooms.

The inter-quartile range can be calculated by a far easier method. The lower quartile is found by taking 25% of 200 = 50 and then adding $\frac{1}{2}$. Thus, for the lower quartile, we wish the mean of the 50th and 51st smallest values in the data set. Both of these values are 3 and so the lower quartile is 3 rooms. The upper quartile is 5 rooms and so the inter-quartile range is 2 rooms. (Note that, as the size of the data set increases, we can begin to ignore the addition of $\frac{1}{2}$.)

If we now consider the data in Table 5.3, we have a slightly different problem. For the range, all we can really say is that the range is less than £49,000.

For the standard deviation and the mean absolute deviation, we reproduce the technique which we used to calculate the mean. Thus, we represent each group by a "typical" value—the mid-point. We now have the data in the same form as the example immediately above and we can evaluate these measures. You should be able to check that the standard deviation is £11,397.27 and the mean absolute deviation is £9,388.75.

For the inter-quartile range, it is best to look at an ogive with percentage cumulative frequency on the vertical axis and to estimate the two quartiles from this. If we read off this graph at 25% and 75%, the answers should be fairly close to 21 and 38 respectively, and so we can estimate the inter-quartile range to be about £17,000.

5.3.9　*Presenting Summaries of Location and Spread*

Traditionally, statistics has been about calculating means and standard deviations for different data sets. While the mean and standard deviation have some desirable properties and allow us to answer some questions about the underlying population, there are some disadvantages. Not only can the calculations be long and tedious but they do not afford us with a clear pictorial method of presentation.

One type of summary of location and spread which does allow us to draw a helpful and informative picture is called a five-number summary (or a five-figure summary) or, alternatively, Tukey's five-number summary. This is due to John Tukey, a prominent statistician who is an exponent of exploratory data analysis as a forerunner to confirmatory data analysis. In other words, he is an exponent of the view that we ought to look at a data set before we begin to ask questions about it.

For a five-number summary, we need to have *six* numbers! We need the number of observations in the data set, their median, the two quartiles, and the two extreme values. Once we have this information, we put it in a box as follows.

$$n =$$

median	
lower	upper
quartile	quartile
smallest	largest
value	value

Thus, in the case of the data on rented property, we have a five-number summary of

$$n = 17$$

	14
9	23
5	74

On its own, a five-number summary is merely a neat way to present a summary of location and spread. However, it is when we have several data sets which we wish to compare and contrast that it becomes particularly useful. Here, we can use the five-number summaries to construct a very useful and informative diagram which is known as a box-and-whisker plot.

5.3.10 Box-and-Whisker Plots

To see how we might use a box-and-whisker plot, consider the following example. Four local authorities were selected in the UK and a check was made on their building programmes. In particular, data were collected on all the houses which reached completion in the last three months and, more importantly, on how long these houses had taken to be completed. A sample of the data appears below with the times recorded to the nearest week.

Authority A: 29 44 51 46 33 53 74 36 48
Authority B: 38 47 63 75 59 76 82 39 53 61 64 76 70 66 61
Authority C: 46 52 47 53 51 50 48 45 50 47 44
Authority D: 21 66 75 82 24 68 55 53 29 76 43 35

With these data, we can easily construct four five-number summaries.

$n = 9$		$n = 14$		$n = 11$		$n = 12$	
	46		63.5		48		54
36	51	53	75	46	51	32	71.5
29	74	38	82	44	53	21	82
A		*B*		*C*		*D*	

For each of the local authorities, we can construct a box-and-whisker. This consists of a box stretching from the lower quartile to the upper quartile with a line across it at the median. There are also two lines stretching out from the box to the extremes.

Now, on its own, a box-and-whisker plot does not do much. However, we can construct a multiple box-and-whisker plot by

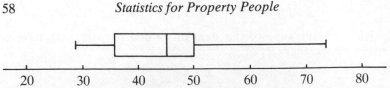

placing the four box-and-whisker plots next to each other on their sides. This multiple box-and-whisker plot allows simple graphical comparison.

Fig 5.1

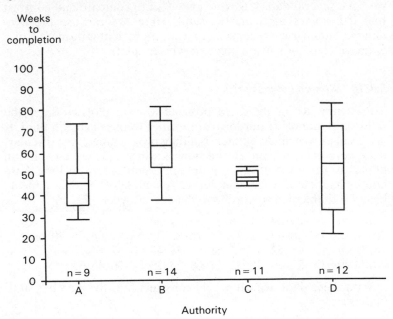

Once the four sets of figures are on the same graph, it is easy to make comparisons. If we remember that the box in the middle contains the middle 50% of the data, we can see that authority *B* tends to take longer, on average than the other authorities whereas authority *A* tends to take less time. As far as time to completion is concerned, there is very little to choose between authorities *C* and *D* but clearly, if you desire reliability or predicability, then authority *C*, with a far smaller spread, is the choice.

Thus, we can see that a multiple box-and-whisker plot is visually potent. In particular, it is useful in many situations where other types of graphs and pictures are inappropriate. Back-to-back histograms and stem-and-leaf displays have the severe limitations that they cannot be used to look at more than two data sets at a time. Also, if there are only two data sets but they are of disparate size, then these two types of graphs can be very misleading. A multiple box-and-whisker plot avoids these problems and, in general, is hard to beat for allowing simple comparisons.

5.4 Symmetry and Skewness

We are often concerned about a data set and whether or not it is symmetric. By a data set being symmetric, we usually mean that the observations are split roughly equally on either side of the mean and the distribution or shape on either side of the mean is similar, if not identical. Thus, if we consider the four histograms below, we would conclude that the first and second histograms are of data sets which we might think of as being symmetric.

The third and fourth histograms are of data sets which most definitely are not symmetric i.e. they are asymmetric. Another term for this is to say that the data sets are skewed. The data set in the third histogram is right skewed or positively skewed.

Fig 5.2

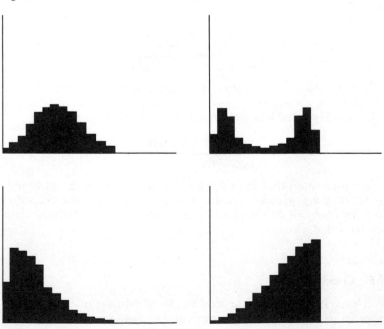

The data set in the fourth histogram is left skewed or negatively skewed. For right skewed data sets, the mean is to the right of the median i.e. the mean is greater than the median. For left skewed data sets, the mean is to the left of the median.

Where do these arise in practice? In reality, the question ought to be "Where can we expect to find data sets which are not skewed?" However, to answer the first question, data on personal income or prices of houses are often quite clearly skewed. In fact, they will tend to be right skewed. In present times in the UK, data on duration of personal spells of unemployment are, unfortunately, likely to be left skewed.

5.4.1 Measures of Skewness

Without going into too many details, there are two measures of skewness which are used more often than others.

The first is a mathematical-type expression which has many desirable properties and is of a similar form to the variance. For the top line, instead of taking the square of the deviations from the mean, this measure takes the cube of these deviations. For the bottom line, this measure has the third power of the standard deviation. This measure of skewness can be written in symbolic terms as

$$\frac{\sum_{i=1}^{n} (x_i - \bar{x})^3}{n} \quad \text{by} \quad \frac{\sum_{i=1}^{n} (x_i - \bar{x})^3}{\sigma^3}$$

Despite the desirable properties mentioned above, we shall not be using this measure in this text. The other measure which is often used is called Pearson's coefficient of skewness and equals

$$\frac{3(\text{mean} - \text{median})}{(\text{standard deviation})}$$

We must note that the measures of skewness do not have any units; they are without a scale. Thus, in measuring the skewness of a data set, we do not need to know the units in which the data are recorded.

5.5 Conclusion

We have now come to the end of the third part of the text and now we should be able to present and summarize data in a useful and informative way. What follows in the rest of the text are specific tools and methods of analysis which will allow us to answer particular questions. One might think that answering questions is far more important than describing a data set but they are equally important. In general, it is only by looking at a data set in an intelligent way that we can begin to pose sensible questions. Further, although we may be able to answer questions by analysing situations or by performing some calculations, it is generally with the help of summaries and diagrams that we present the results of our investigations and win over our colleagues and our competitors to our point of view. Without this ability to win over these people, we will not be able to properly use and implement our statistical knowledge. Therefore, it is important that the contents of this chapter are not only understood but become part of the set of techniques which you call on daily as you tackle problems in your professional lives.

Problems

1. Nine recent house sales in Leeds were for the following prices:

£27,100	£36,400	£29,150	£58,500	£32,450
£41,000	£53,000	£38,230	£34,850	

    Calculate the mean, median, and mode of these prices.
    Calculate the range, mean absolute deviation, inter-quartile range, and standard deviation of these prices. On the basis of this evidence, are house prices in Leeds skewed to the right or to the left or roughly symmetric?

2. A sample of 43 companies' share prices are as follows:

127	98	248	334	95	79	49	27	125	143
375	44	31	60	265	218	859	62	609	124
240	377	238	505	8	255	200	153	12	52
630	21	445	100	174	95	46	320	196	340
441	286	98							

    Calculate the mean, median, and mode share price.
    Construct a five-figure summary of these share prices.
    Calculate the range, inter-quartile range and standard deviation of these share prices.

3. The rateable values of 250 properties are as follows:

Rateable Value (£)	Number of Properties
100–199	3
200–299	7
300–399	41
400–499	65
500–599	61
600–699	54
700–799	17
800–899	2

Construct a cumulative frequency table of these rateable values. Using this, draw the corresponding ogive and use this to estimate the median rateable value and the 29%-ile.
Estimate the mean and the standard deviation.

4. Five competing estate agents have called in an independent consultant. He has been instructed to investigate the advantages and disadvantages of these four estate agents amalgamating into one large estate agency. As part of his study, he has collected information on the operations of each of the individual estate agents. One piece of information which he has collected is the number of telephone enquiries each estate agent receives in a period of an hour on a Wednesday. (Wednesday is the main day for domestic property in this area.) The numbers of enquiries for each estate agent for a number of hour-long periods are as follows:

Agent A
23 10 16 17 28 41 10 18 25 22 19 15 20
Agent B
19 16 14 11 32 30 14 17 29 27 20 31 23 26
Agent C
18 45 28 25 18   9 23 25 40 17
Agent D
16 25 18 30 33 15 24 28
Agent E
14 22 29 34 27 25 21 43 41 28 24 26 28 22 25

For each of the agents, construct a five-figure summary of the number of enquiries which they received.
For agent D, calculate the mean absolute deviation and the variance of the number of enquiries.
For agent E, calculate the mean number of enquiries and the standard deviation of the number of enquiries.
Use the five-figure summaries to construct a multiple box-and-whisker plot. Briefly comment on whether the distributions of

the number of enquiries which each agent receives appears to be similar. Identify any differences which might be of importance to the consultant.

5. Two cities are comparing their recent records of selling sites for development. The areas of their recent sales are recorded in hectares

City 1	1.31	1.63	1.65	0.94	1.32	1.45	1.73	2.37
	1.57	2.24	1.83	1.93	0.55	2.82	2.45	
City 2	1.04	1.36	1.94	2.26	2.39	0.48	0.82	2.14
	1.28	1.17	2.03	0.68	0.34	0.85	1.46	1.68
	1.33	1.62						

For city 1, calculate the mean and median area of their recent sales and briefly comment on the symmetry.

For city 2, calculate the standard deviation and the inter-quartile range of these areas.

For the two data sets, construct a back-to-back stem-and-leaf display and a multiple box-and-whisker plot.

PART FOUR
Probability

CHAPTER 6

Probability

6.1 Introduction

The subject of probability is simultaneously theoretical and practical. On one hand, probability involves some difficult concepts while, on the other hand, probability is the basic tool of statistics which allows us to actually provide answers to some, if not all, of the questions which we might ask.

A number of books on the applications of Statistics omit to have a section on probability. However, such an omission hinders the reader from either gaining a good intuitive feel for probability and, therefore, for the results of statistical analyses, or extending standard techniques so that the reader might be able to cope with problems which are not exactly of a standard type. Therefore, although the following discussion of probability may seem a little theoretical, unrelated, and a little out of place, rewards for a good understanding of probability will be reaped in chapters to come.

6.2 Some Definitions

Before we look at some definitions, let us consider the example in chapter 5 on the data collected on the number of rooms in a house. There were two hundred families and suppose that we are going to select one of these families so that we can ask them a further series of questions. What is the chance that we pick a family whose house has six rooms?

If we have a method of selection which gives each family the same chance of being selected as any other, then the answer that most people would give to the above question is $\frac{28}{200} = 0.14$.

(Note here that some selection methods do *not* give each family the same chance of being selected. Some of these will be discussed in the section on Sampling. However, for the moment, we can consider that the names of the 200 families were put on 200 pieces of paper which were rolled up and shaken well in a hat before one of them was picked out.)

Now, we can discuss all the possible numbers of rooms we can get when picking one of these families. These are the numbers 1, 2, 3, 4, 5, 6, 7, 8 and 11. Each of these possible answers is called a sample point and the collection of them is called a sample space.

Definition: We shall loosely define a trial or an experiment as the process of performing some action and looking at the result.

Definition: A sample space is the set of all possible outcomes of an experiment or trial.

Definition: Each one of these possible outcomes which together comprise the sample space is called a sample point.

Suppose we are looking at overcrowding in houses. Thus, we might only be interested in the families with 3 or fewer rooms. We can call this an event. Thus, the event 'family has 3 or fewer rooms' is represented in the sample space by the sample points 1, 2 and 3.

Definition: A collection of sample points with a common property is called an event.

As regards the answer to the question "What do we mean by probability?", there are many answers. Philosophically, there are at least four distinct views of what we mean by saying that the probability that something happens is 0.4. However, we shall look at just one of these—the relative frequency approach to probability.

By saying that the chance of picking a family with six rooms is 0.14, what we are really saying is that, if we were to repeat the trial several hundred times (or even more), then about 14% of the trials would have a family with six rooms.

Definition: The probability of a sample point is the proportion of occurrences of the sample point in a long run of experiments.

In theoretical terms, this long run of experiments is infinitely long.

The notion that we shall use here is that the probability that the sample point x occurs is denoted by $P(x)$.

By thinking about this relative frequency approach, we can see quite clearly that the probability of a sample point occurring can never be less than 0, for the number of occurrences can never be negative. Also, the probability can never exceed 1, for the number of times a sample point occurred can never be greater than the number of trials which were carried out.

If the sample point always occurs—probability is 1.
If the sample point never occurs—probability is 0.

Thus, the minimum probability that we can ever envisage is 0 and the maximum is 1. Further, given that the proportion of occurrence of each of the sample points must add up to 100%, we can say that the sum of the probabilities of all the sample points is 1.

Given this fairly unrestrictive definition of probability, we can now progress to some rules and definitions. These can, in general, be called probability theory by which we mean a set of rules, definitions, etc. for manipulating probabilities and for calculating the probabilities of simple and complex events.

Definition: The probability of an event happening is the sum of the probabilities of the individual sample points which together have the common property which defines the event.

Thus, in symbolic terms we can write that

$$P(\text{event}) = \Sigma P(\text{sample points in the event})$$

We shall consider some simple and, unfortunately, unrealistic examples. These deal with equally likely sample points and so the probabilities of the sample points occurring are the same for all the sample points.

Example. Toss two ordinary coins—look at the uppermost faces. (*H* denotes a "head", *T* denotes a "tail")

Sample space	HH	HT	TH	TT
Probability	0.25	0.25	0.25	0.25

The event 'at least one tail appeared' corresponds to the sample points *HT, TH, TT.* Thus, the probability that at least one tail appeared is $0.25 + 0.25 + 0.25 = 0.75$ (vid definition above).

Example. Roll an ordinary die twice—look at the uppermost faces.
The sample space contains 36 sample points. [(3,5) denotes a '3' on the first roll and a '5' on the second].

(1,1)	(1,2)	(1,3)	(1,4)	(1,5)	(1,6)
(2,1)	(2,2)	(2,3)	(2,4)	(2,5)	(2,6)
(3,1)	(3,2)	(3,3)	(3,4)	(3,5)	(3,6)
(4,1)	(4,2)	(4,3)	(4,4)	(4,5)	(4,6)
(5,1)	(5,2)	(5,3)	(5,4)	(5,5)	(5,6)
(6,1)	(6,2)	(6,3)	(6,4)	(6,5)	(6,6)

I have said that these sample points are all equally likely to happen and so the probability that each one of them occurs is $\frac{1}{36}$.

What is the probability associated with the event "sum of the numbers on the two faces is equal to 5"? To get this, we need to identify the sample points associated with this event—(1,4), (2,3), (3,2) and (4,1)—and add up the probabilities associated with these sample points $= \frac{4}{36} = \frac{1}{9}$.

What is the probability that the number on the first roll of the die is larger than the number on the second roll? We first of all need to identify the sample points associated with this event—there are 15 in all—and then add up their probabilities. This equals $\frac{15}{36} = \frac{5}{12}$.

In chapter 4, we dealt with verbal quantifiers. We shall be using them frequently with respect to probability. If you are not sure of the quantifiers, go back and read 4.3.

We can now use two of these quantifiers as we progress to the first of the two basic rules for manipulating probabilities.

Addition Rule for Probability

Let A and B be any two events in which we may be interested.

$$P(A \text{ or } B) = P(A) + P(B) - P(A \text{ and } B)$$

As an example of how this works, we can solve a problem asked above. If we toss a coin twice, what is the probability that a tail appears at least once?

We can restate this as the probability that there is a tail on the first toss or there is a tail on the second toss and we can now solve that using the Addition Rule.

Let A be the event "there is a tail on the first toss", and
Let B be the event "there is a tail on the second toss".

Using the rule, we can now say that

$$P(A \text{ or } B) = P(A) + P(B) - P(A \text{ and } B).$$

From looking at this problem earlier, we know all the probabilities on the right hand side of the equation.

Thus, $P(A \text{ or } B) = \frac{1}{2} + \frac{1}{2} - \frac{1}{4} = \frac{3}{4}$.

Example. Roll an ordinary die twice and look at the numbers on the uppermost faces. What is the probability that we get at least one (3) or a sum of 7?

Let H be the event "there is at least one (3)" and
Let S be the event "the sum is 7".

The question asks for the probability of either H or S happening. We can write this as $P(H$ or $S)$ and we know that this equals $P(H) + P(S) - P(H$ and $S)$. We can, once again, evaluate all the terms on the right hand side.

There are 11 sample points associated with H. Thus, $P(H) = \frac{11}{36}$. There are 6 sample points associated with S. Thus, $P(S) = \frac{6}{36}$. There are two sample points which are associated with both H and S. These are the sample points $(3, 4)$ and $(4, 3)$. Thus, $P(H$ and $S)$ equals $\frac{2}{36}$. Thus, the $P(H$ or $S) = \frac{11}{36} + \frac{6}{36} - \frac{2}{36}$ which equals $\frac{15}{36}$ or $\frac{5}{12}$.

6.3   Types of Events

Definition: Two events are said to be mutually exclusive if they have no sample points in common. Thus, in a single trial, mutually exclusive events cannot both occur.

Example. Roll a die—look at the number on the uppermost face.
Consider the following events.

E_1: number is divisible by 5
E_2: number is even
E_3: number is divisible by 3

Events E_1 and E_2 are mutually exclusive. They have no sample points in common. The sample point associated with the former event is (5) while the sample points associated with the latter event are (2), (4) and (6). Thus, they do not have any sample points in common. What this means is that, when rolling an ordinary die, it is impossible to get a number on the uppermost face which is both even and divisible by 5. (Such numbers do exist but are not on the faces of ordinary dice.)

Similarly, E_1 and E_3 are mutually exclusive because there is not a number on the face of an ordinary die which satisfies both of these conditions—divisible by 5, and divisible by 3—at the same time.

On the other hand, E_2 and E_3 are *not* mutually exclusive. They do have a sample point in common—(6)—and so it is possible, if there is a (6) on the uppermost face, for both of these events to occur at the same time.

The Addition Rule as stated above is always true but, if the events in which we are interested are mutually exclusive, then the Rule can be stated in a simpler way. This is because the events are mutually exclusive and so the probability of them both occurring is zero, i.e. $P(A$ and $B) = 0$. Thus, we get

Addition Rule for Probability for Mutually Exclusive Events

Let C and D be two events in which we are interested and let them be mutually exclusive.

$$P(C \text{ or } D) = P(C) + P(D)$$

You can try some examples of this specialized version of the rule with the die-rolling situation discussed when we introduced the idea of mutually exclusive events.

Definition: Two events, A and B, are said to be complementary if

(i) they are mutually exclusive
and
(ii) together, they comprise the sample space.

In terms of probability, we can define complementary to be such that

(i) $P(A \text{ and } B) = 0$
and
(ii) $P(A) + P(B) = 1$

In non-mathematical terms, we can see that if two events are complementary, exactly one of them occurs at any one time. Thus, the occurrence of one of them implies the non-occurrence of the other and vice versa.

For example, if we consider people, then, in common terms, we could define the following pairs of events to be complementary: male/female, employed/unemployed, under-18/over-18, owns shares/does not own shares, etc.

In notational terms, we often denote the event which is complementary to A by $\acute{A}$.

However, we can immediately extend the idea of complementary events to the mathematical idea of a partition. We can define a partition to be a set of 2 more events which satisfy

(i) each pair of events is mutually exclusive
and
(ii) all the events together comprise the sample space.

Once again, if we consider people, we might define a partition of under-18/18–35/36–59/over-60, upper-class/middle-class/lower-class, or Caucasian/Mongoloid/Negroid.

6.4 Conditional Probability

Suppose we are playing a board game where we need to throw a "6" to start. We may be interested in the probability that we throw a number which is 3 or less. This probability is equal to $\frac{3}{6}$ or $\frac{1}{2}$. However, it is the beginning of a game and a friend is watching. This friend did not see your roll of the die but can see from the board that you have not yet started. What now is the probability that we threw a 3 or less? Well, originally we had six possible outcomes to the roll of the die. However, as we have not started we know that a "6" is ruled out. Whatever number we got, it was not a "6". We now only have five possible outcomes. These are still equally likely and, to ensure a sum of 1, must each have a probability of $\frac{1}{5}$. Thus, the probability that we threw a 3 or less is $\frac{1}{5} + \frac{1}{5} + \frac{1}{5} = \frac{3}{5}$. Previously, the probability was $\frac{1}{2}$; now it is $\frac{3}{5}$. What has happened to change this probability?

What has happened is that we now are dealing with a conditional probability. We have been given some information about what has happened and any probability we consider must take into account this information.

We might estimate that you arrive at work on time on 95% of workdays. Thus, the probability that, on any chosen day, you arrive on time at work is 0.95. However, this probability would alter if we know that all public transport workers are on strike and so the roads are very busy. You may still be on time but one would expect the probability of being on time to drop. Similarly, if we know that your alarm clock failed to ring, the probability of you arriving at work on time would be expected to drop.

In technical terms, the information contained either restricts the sample space, as in the die example above, or it alters the probabilities of the sample points.

If we have A as the event of interest and B is the event which we are told has occurred, then we use the notation $P(A/B)$ to mean the probability that A occurs if we are told that B has occurred. In a shorter form, we can often read $P(A/B)$ as "the probability of A given that B has occurred" or, even shorter, "the probability of A given B". Note that this is *not* the division of A by B.

Example. In a particular town, 35% of the bungalows have adjoining garages. However, in an estate called East Wilton, which was built at a time of high street vandalism, 54% of the bungalows have adjoining garages. I am considering buying a specific bungalow in this town but have not yet been to see it and I do not know if it has an adjoining garage. What is the probability that it has such a garage? If I now find out that the bungalow is in East Wilton, what now is the probability that there is an adjoining garage?

Let us write
A: bungalow has an adjoining garage.
B: bungalow is in East Wilton.

The answer to the first part is clearly $P(A)$ and the question tells us that the answer is 0.35.

However, the answer to the second part is the probability that the bungalow has an adjoining garage if we are told that it is in East Wilton. In terms of the events, the question is asking for the probability that event *A* occurs if we are told that event *B* occurs and we can write this as $P(A/B)$. The answer to this is given in the question as 0.54.

We can now progress to the second of the two basic rules for manipulating probabilities.

Multiplication Rule for Probability

Let *A* and *B* be any two events in which we may be interested. Then

$$P(A \text{ and } B) = P(A/B) \times P(B)$$

and

$$P(A \text{ and } B) = P(B/A) \times P(A)$$

Example. Suppose, in the above example, that 20% of the town's bungalows are, in fact, in the East Wilton estate. We could now ask "What is the probability that the bungalow which is being considered has an adjoining garage and is in the East Wilton estate?"

What we are really asking for in symbolic terms is $P(A$ and $B)$. Using the Multiplication Rule, we can now evaluate this as we know that $P(B) = 0.2$ and that $P(A/B) = 0.54$. Thus, we can work out that $(0.2 \times 0.54 = 0.108)$ 10.8% of the town's bungalows are situated in East Wilton with adjoining garages.

Example. What is the probability of drawing two aces if we draw two cards simultaneously from a standard set of 52 playing cards?

Let   A: ace on the first card
and let   B: ace on the second card.

The question is really asking for $P(A$ and $B)$ which, by the above rule, equals $P(B/A) \times P(A)$. There are four aces in a pack and so $P(A) = \frac{4}{52} = \frac{1}{13}$. For $P(B/A)$, we need to think about the situation and see that, if A has occurred, then there are only three aces left in a pack of 51. Thus, $P(B/A) = \frac{3}{51} = \frac{1}{17}$. The answer to the problem is $\frac{1}{13} \times \frac{1}{17} = \frac{1}{221}$.

Definition: Two events, A and B, are said to be statistically independent if knowledge about whether one of them has occurred does not affect the probability of the other one of them occurring.

If two events are not statistically independent, then we say that they are statistically dependent. (Note that, in general, we drop the qualifier "statistically".)

In the above example on bungalows with adjoining garages, $P(A) = 0.35$ while $P(A/B) = 0.54$. Thus, knowledge that the bungalow is in East Wilton alters the probability that the bungalow has an adjoining garage. Thus, the two events A and B are dependent. This is not to say that one causes the other. It merely says that knowledge of one of them occurring gives us some feeling about the probability of the other one occurring.

6.5 Two Basic Theorems

While the Addition Rule and the Multiplication Rule are fundamental rules, we need to establish two theorems which will really open up the whole area of probability. While I shall develop these theorems from first principles, we already have covered the necessary concepts to state and prove them.

Let us suppose that the town mentioned above has bungalows in four estates and nowhere else: East Wilton, Highville, Westerbury, and Carrowburn. Thus, we can consider these four estates to form a partition in the sense that all bungalows in this town are in one and only one of these estates.

Suppose that we want to know the probability that a selected bungalow has an adjoining garage. Call this event A. If we define some more events, the theorem will be fairly simple to state and prove. Let B_1 denote "bungalow is in East Wilton". Similarly, let B_2, B_3, and B_4 denote the events that the bungalow is in Highville, Westerbury, and Carrowburn, respectively.

Let us suppose that we are interested in the probability that a bungalow has an adjoining garage, i.e. $P(A)$.

We can write $P(A)$ as $P(A$ and B_1 or A and B_2 or A and B_3 or A and $B_4)$.

All this says is that a bungalow with an adjoining garage is either a bungalow with an adjoining garage in East Wilton or is a bungalow with an adjoining garage in Highville, etc. Now, we can assume that these four estates are non-overlapping, i.e. the events B_1 to B_4 are mutually exclusive and we have already seen that, if events F and G are mutually exclusive, $P(F$ or $G)$ equals $P(F) + P(G)$. Thus, we split up the above probability under consideration into four parts. Thus, $P(A)$ can be written $P(A$ and $B_1) + P(A$ and $B_2) + P(A$ and $B_3) + P(A$ and $B_4)$.

Now, if we use the Multiplication Rule, we can rewrite each of these four terms and so we can now write $P(A)$ as

$$P(A/B_1) \times P(B_1) + P(A/B_2) \times P(B_2)$$
$$+ P(A/B_3) \times P(B_3) + P(A/B_4) \times P(B_4)$$

This is a statement of the Theorem of Total Probability.

More formally, we can define the Theorem of Total Probability as follows:

If $B_1, B_2, \ldots B_n$ form a partition, and we are interested in the probability of an event A, then we can write

$$P(A) = \sum_{i=1}^{n} [P(A/B_i) \times P(B_i)]$$

To try to make this clearer, we can associate numbers with some of these events. Suppose that there are 10,000 bungalows in this town, 4,000 in East Wilton, 3,000 in Highville, 2,000 in Carrowburn and 1,000 in Westerbury. Further, suppose that 54% of the bungalows in East Wilton have adjoining garages, and that this percentage is 28% in Highville, 31% in Carrowburn, and 43% in Westerbury. What percentage of all bungalows in this town have adjoining garages?

Well, 54% of the 4,000 in East Wilton have adjoining garages. That is 2,080 bungalows. We also have 28% of 3,000, which is 840, 31% of 2,000, which is 620 and 43% of 1,000, which is 430. This gives a total of $2,080 + 840 + 620 + 430 = 3,970$, out of 10,000, or 0.397.

The theorem says that

$$P(A) = P(A/B_1)P(B_1) + P(A/B_2)P(B_2) + P(A/B_3)P(B_3)$$
$$+ P(A/B_4)P(B_4)$$
$$= 0.54 \times 0.4 + 0.28 \times 0.3 + 0.31 \times 0.2 + 0.43 \times 0.1$$
$$= 0.208 + 0.084 + 0.062 + 0.043$$
$$= 0.397$$

Thus, it seems to correspond with the arithmetic procedure we might carry out if we had the actual numbers.

The other theorem which we shall need is due to an 18th century

English minister called Thomas Bayes. It can be developed very easily from the Multiplication Rule.

As we saw, $P(A \text{ and } B) = P(A/B) \times P(B)$
but we also have that $P(A \text{ and } B) = P(B/A) \times P(A)$

Thus, the two expressions on the right hand side of these equations must be equal, i.e.

$$P(B/A) \times P(A) = P(A/B) \times P(B)$$

and we divide both sides by $P(A)$, we arrive at a statement of Bayes' Theorem.

Bayes' Theorem:

$$P(B/A) = \frac{P(A/B) \times P(B)}{P(A)}$$

To see how to use Bayes' Theorem and the Theorem of Total Probability, let us look at an example.

Example. Depending upon how well-designed buildings are, they are classified by the West Macao Development Corporation as being either stable or wobbly. It is reckoned that, if a building is stable, the probability of it cracking when there is an earth tremor of greater than 5.0 on the Richter Scale is 0.4, while, for unstable buildings, the probability is 0.9.

About 80% of the buildings in the region are, thankfully, classified as being stable. However, the chairman of the development corporation is not sure about his own house. The house comes with the job and so he never found out if it was stable.

On June 20th 1981, during the midsummer fertility rites, there was an earth tremor which was recorded at over 5.0 on the Richter Scale. When the chairman returned to his house, he discovered that it had cracked. What is the probability that his house is classified as being stable?

If we define the events

A: cracked under an earth tremor of greater than 5.0
B: property is classified as being stable
   then the question is really asking for $P(B/A)$.
   By Bayes' Theorem, we see that this equals

$$\frac{P(A/B) \times P(B)}{P(A)} \quad \text{which equals} \quad \frac{0.4 \times 0.8}{P(A)}$$

Now, we need to evaluate $P(A)$ and we can use the Theorem of Total Probability to evaluate this. We can do this by using the fact that "stable" and "wobbly" are complementary and, therefore, form a partition. Thus, if we denote wobbly by B', we can write $P(A)$ as

$$P(A) = P(A/B) \times P(B) + P(A/B') \times P(B')$$

and we can evaluate this as $0.4 \times 0.8 + 0.9 \times 0.2 = 0.5$.
Thus, the answer which the chairman is seeking is $0.32/0.5 = 0.64$.

Now, you may wonder why and where we practically use Bayes' Theorem. Well, we really use it all the time. We would call $P(B)$ the prior probability that the house is stable while we would call $P(B/A)$ the posterior probability. Thus, we adjust our prior probability in the light of new information and we could do this for further pieces of information. In this way, we continually update our probability as and when we get more information. Do we ever do this in life? Well, we all do it every day. Consider a first meeting with a business associate. Immediately, you form an opinion, either on the basis of appearance or on the basis of a form of stereotype. As you deal with this person, you find out more information about them and continually adjust your opinion of their personality, ability, etc. We tend to do this subconsciously and very quickly and not always particularly rationally but we are updating opinions in the same way as Bayes' Theorem does; for it merely updates our opinion of the probability of an event happening. Thus, in statistics, it is a well-known saying about this subject that "Today's posterior is tomorrow's prior."

Is this sort of thing done professionally? Suppose a colleague asked for your opinion of the value of a plot of land. What would you do? Presumably, you would begin to ask a series of questions which would allow you to develop a posterior assessment of the plot's worth. Each new piece of information will cause the assessment to change. What questions would you ask? Well, the first few might be "Where is it?", "What size is it?", "What is it to be used for?", "What is it used for now?", "What is the situation as regards planning permission?", "Are we talking about a price to buy it for or a price to sell it for?", and "Whose is it now?". The answer to each of these questions would allow you to adjust your assessment and, eventually, to offer a reasonably accurate answer.

At this point, we can leave the theory of probability as we have really covered all we need for a reasonable understanding and to allow us to consider probability distributions in the chapters to follow. Once again, it must be stressed that an understanding of probability is vital for anyone involved either in performing or in interpreting statistical analyses. Thus, if you are not clear about some of the ideas in this chapter, *stop now* and go back and read the chapter again and try some examples.

Problems

1. In a factory, 60% of the staff are male. It is also known that 70% of the staff are 35 years of age or older. Further, 42% of the staff are both male and 35 or older. What percentage of the staff is either male or 35 years of age or older? What percentage of the staff is both female and under 35?

2. Of recent loans given by a building society, 0.08 were for properties in Scotland and 28% were for properties built before 1940. If, in fact, 60% of the loans for Scottish properties were for properties built before 1940, what proportion of the building society's recent loans were for properties which are either in Scotland or were built before 1940?

3. There are 800 houses in a new housing estate. There are 8 master keys for these houses—each key opens 100 houses. Further, it is estimated that each house has a 0.3 probability of being left unlocked by the previous potential customer. The site manager is about to show a potential customer a house but has only three master keys in his pocket. If we assume that his possession or not of the master key is independent of whether the house is locked or unlocked, what is the probability that he is able to let the potential customer into the house?

4. A Bristol car showroom has found that 0.3 of the new cars which it sells are foreign. Further, 0.6 of the foreign cars which it sells need new exhaust systems within two years. What is the probability of a car sold by this car showroom being both foreign and needing a new exhaust system within two years? If 0.41 of the British cars need a new exhaust system within two years, what is the probability of a car sold by this showroom being both British and needing a new exhaust system within two years? What proportion of the cars sold by this showroom need a new exhaust system within two years? Are the events "car is foreign" and "car needs a new exhaust system within two years" statistically independent?

5. In a television quiz programme, prizes are locked in boxes. There are 10 boxes, 3 of which contain "booby" prizes and 7 of which contain "good" prizes. Once a box is opened, it is eliminated, and each contestant opens one box. If there are two contestants, what is the probability that neither of them gets a "booby" prize? If there are three contestants, what is

the probability that none of them get a "booby" prize? What if there are four contestants or five contestants?

6. From a study of recent building trends in an area in Wales, it has been found that, of those housing projects completed on time, 80% are privately funded and 20% are funded by public authorities. Of those projects completed late, the split is 40%—private and 60%—public. Further, $\frac{2}{3}$ of all housing projects are completed on time. A private project is about to be started. What is the probability that it will be completed on time? Are the events "on time" and "privately funded" statistically independent?

7. An insurance salesman has found that 20% of all his house calls have resulted in sales. Of those families that buy insurance policies, 35% live in detached homes. Of his unsuccessful calls, 60% were to detached homes. What is the probability that his next call will result in a sale if the call is to a detached home? What is the probability that his next call is unsuccessful if it is to a home which is not detached?

CHAPTER 7

Discrete Random Variables

7.1 Introduction

Previously, we have discussed sample spaces and they can be of two types. Either they are qualitative, for example, colour of hair, type of house, bankrupt/solvent, or they are quantitative, for example, number of rooms, amount of mortgage payments, interest rate.

When the sample space relates to a quantitative characteristic, a number can be associated with each point in the sample space. The following example is of a quantitative random variable. We shall also, in this chapter, meet some qualitative random variables.

Suppose that we consider going to see a property. On inspection, the number of rooms can be counted. Suppose that X denotes the number of rooms. Let's say that, in this example, X can take the values 1, 2, 3, 4, 5, 6 and 7. For any property, X takes one of these values. Which one it takes is, however, uncertain as it will change as we go from property to property. For this reason, X is called a random variable.

As much as is possible, a distinction shall be drawn between a random variable and the value which it takes at any particular time. A capital letter shall be used to denote the random variable and a lower case letter shall be used to denote the value of the random variable.

Definition: A discrete random variable is a random variable which takes only discrete/distinct values.

Examples of discrete random variables are:
The number of employees in a surveyor's office.
The number of houses built on a housing estate.
The number of changes of interest rate in a year.

Definition: A continuous random variable is a random variable which takes values on a continuous scale.

Examples of continuous random variables are:
The dimensions of a shop front.

The temperature in the rooms of a house.
The area of a garden.

As was mentioned earlier, it should be noted that the above three variables are often measured and recorded in discrete units. This is because of the recording instruments which are used and the accuracy which is available to us. However, they are inherently continuous.

We can now proceed to discuss discrete probability distributions. Continuous probability distributions will be dealt with later on.

7.2 Discrete Probability Distributions

Definition: The probability distribution for a discrete random variable X associates with each distinct outcome x_i, a probability $P(X = x_i)$. Moreover, these probabilities are such that

$$0 \leqslant P(X = x_i) \leqslant 1 \qquad \text{for } i = 1, 2, \ldots, k$$

and

$$\sum_{i=1}^{k} P(X = x_i) = 1$$

As an example of a discrete probability distribution, we can consider the case of the number of rooms in the houses on a housing estate. Let X denote the number of rooms in a house. The probability distribution might be

x:	1	2	3	4	5	6	7
$P(X = x)$:	0.05	0.1	0.15	0.2	0.3	0.15	0.05

I.e. the probability of a selected house having 4 rooms is 0.2, and the probability of a selected house having 6 rooms is 0.15. This can be represented graphically (see Figure 7.1).

7.3 Cumulative Probability Distributions

Sometimes we are interested in the probability of a random variable X taking a value less than or equal to some value x. This set of probabilities is called a cumulative probability distribution and is denoted $P(X \leqslant x)$. A cumulative probability distribution gives values of $P(X \leqslant x)$ for all values of x.

In the above case, the cumulative probability distribution would be

x:	1	2	3	4	5	6	7
$P(X = x)$:	0.05	0.15	0.3	0.5	0.8	0.95	1.0

Fig 7.1

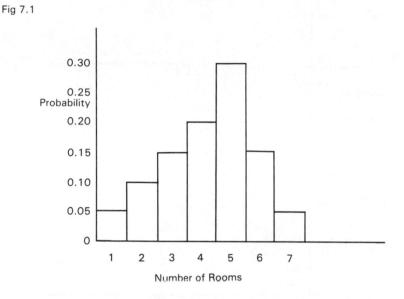

I.e. the probability of a house having 3 or fewer rooms is 0.3—this corresponds to $P(X = 1) + P(X = 2) + P(X = 3)$. By considering complementary events, we can also get the probability that a selected house has 6 or more rooms. The event "6 or more rooms" is the complement of the event "5 or fewer rooms". Thus, the required probability must equal $1 - 0.8$ which equals 0.2.

It was said above that this cumulative probability distribution is defined for all values of x. Thus, despite the fact that, in this example, it makes little sense, we can answer the question "What is the probability that the number of rooms is 3.6 or less?" Clearly, for X to take a value less than or equal to 3.6, it must take a value less than or equal to 3 as no value between 3 and 3.6 is possible. Thus, $P(X \leqslant 3.6) = 0.3$.

In fact, to make this idea clearer, we can draw a graph of x against $P(X \leqslant x)$. For any value of x, the cumulative probability can be read off this graph (Figure 7.2).

7.3.1 Bivariate Discrete Probability Distributions

Looking at one variable (or at one quantity) at a time is often of little use. In fact, what is usually of interest is the way in which variables (or quantities) interact with each other. We can look at two variables at a time in a bivariate probability distribution.

Fig 7.2

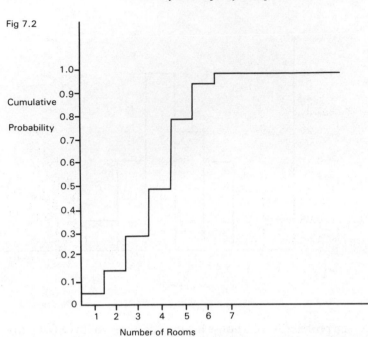

Number of Rooms

Definition: A bivariate probability distribution for two discrete random variables, X and Y, associates with each pair of outcomes, x_i and y_j, a probability $P(X = x_i$ and $Y = y_j)$. These probabilities are, as before, subject to the same two conditions:

$$0 \leqslant P(X = x_i \text{ and } Y = y_j) \leqslant 1 \qquad i = 1, 2, \ldots, k \qquad j = 1, 2, \ldots, m$$

and

$$\sum_{i=1}^{k} \sum_{j=1}^{m} P(X = x_i \text{ and } Y = y_j) = 1$$

To make this idea a bit clearer, we can look at an example. A local authority have recently been investigating types of properties which home-owners have before and after moving home. Properties were defined to be of one of four types: F (flat), T (terraced), SD (semi-detached), and D (detached). They recorded what type of house each home-owner had before they sold and what type of house they bought. These two are denoted by Before and After in the table below. The entries are the proportions in the investigation which fell into each category. As the investigation was quite large, these proportions can be thought of as being probabilities. (Note that these are qualitative variables.)

B		AFTER			
		F	*T*	*SD*	*D*
E	*F*	0.09	0.06	0.11	0.02
F	*T*	0.01	0.05	0.13	0.06
O	*SD*	0.01	0.02	0.06	0.11
R	*D*	0.02	0.01	0.03	0.21
E					

Thus, what this table says is that, of the people in the study, 13% of them were moving from a terraced house to a semi-detached house. This is a joint probability, i.e. the probability of someone living in a terraced house before their move and buying a semi-detached house is 0.13. Check that the total of these probabilities is 1.0.

7.3.2 Marginal Probability Distributions

Suppose that we wished to know what proportion of the people in the study started out living in a flat. We can easily do this by adding across the rows in the above table. (We can see that this makes sense from the Theorem of Total Probability.) The resulting probability distribution is

BEFORE	*F*	*T*	*SD*	*D*
	0.28	0.25	0.20	0.27

Of course, this must add up to 1.0. However, it must be noted that we are not saying here that 28% of families live in flats. What is being said is that 28% of the families who moved home in the period of the study lived in flats. It may be that people who live in flats move more frequently or less frequently than people who live in detached houses. This question of what proportion of people live in flats cannot be answered by this study.

Similarly, we can get the marginal distribution for the types of property after the move.

AFTER	*F*	*T*	*SD*	*D*
	0.13	0.14	0.33	0.4

Can we get conditional probabilities of interest from this table? Well, suppose that we know that one family, in particular, lived in a semi-detached house prior to moving. We can ask "What is the probability that, after moving, the family lives in a flat, or a terraced property?"

Well, we can approach this statistically. We are asking for

$$P(F \text{ after}/SD \text{ before})$$

and this equals, by rearranging the probabilities in the multiplication rule,

$$\frac{P(SD \text{ before and } F \text{ after})}{P(SD \text{ before})} = \frac{0.01}{0.20} = 0.05$$

Thus, we have an answer of 0.05 and what, in fact, we have done, is to take the probabilities of the "*SD* before" row and scale them up so that they add up to 1.0, and, therefore, these probabilities constitute a valid probability distribution.

Similarly, for the conditional probability of the family living in a terraced property, we can either think of following the statistical approach or think merely of scaling up the probabilities. Either way, the calculations are identical and the answer is $0.02/0.2 = 0.1$.

7.4   Properties and Functions of Random Variables

Random variables, or variables for short, are very useful. They provide models for certain situations and also, by providing a notation and system for considering problems, allow us to tackle larger problems than we would have been able to with merely the basic ideas of probability which were developed in the last chapter.

Random variables can be treated mathematically and we shall briefly do so for the purposes of setting up a few general rules and formulas for answering some interesting and far-reaching questions. However, it should be noted that most of the functions and properties mentioned below are valid only for quantitative variables.

7.4.1   Mean of a Random Variable

If we can again consider the example of 7.2 which looked at the number of rooms per house for a housing estate. Recall that the random variable X denotes the number of rooms in a house.

$$
\begin{array}{lccccccc}
x: & 1 & 2 & 3 & 4 & 5 & 6 & 7 \\
P(X=x): & 0.05 & 0.1 & 0.15 & 0.2 & 0.3 & 0.15 & 0.05
\end{array}
$$

One question which may be of interest is "What is the average number of rooms?" Well, if we first consider the mode, we either take a sample and use the mode of the sample or we extract a theoretical mode from the probability distribution. This will correspond to the sample point with the largest probability which is, in this case, 5 rooms.

As regards the median, we can look at figure 7.1 and identify the median as 4.5 rooms.

However, it is likely, in most cases, that this question would be asking for the mean. The mean can be quite easily computed from a sample but, as with the mode, we do not need to take a sample of these houses to find the mean. We can compute a theoretical mean from the probability distribution. This mean is often called "the expected value", and is denoted $E(X)$. Thus, as we are dealing with a random variable X, we are going to evaluate the expected value of X: $E(X)$.

If we had taken a sample, then we would have calculated the mean as

$$\left[\sum_{i=1}^{n} x_i f(x_i)\right]\Big/ n$$

However, as we have taken probability to be the relative frequency of an event in a number of trials, we can divide each of the frequencies by n instead of dividing by n after evaluating what is inside the brackets. This sounds as if it would take longer but, in fact, what it means is that the frequencies divided by n now become probabilities. Thus, we now have a formula for evaluating expected values.

$$E(X) = \sum_{i=1}^{n} x_i P(X = x_i)$$

Thus, in the example above, we can evaluate the expected value or the mean to be

$1 \times 0.05 + 2 \times 0.1 + 3 \times 0.15 + 4 \times 0.2 + 5 \times 0.3 + 6 \times 0.15 + 7 \times 0.05$ which equals $0.05 + 0.2 + 0.45 + 0.8 + 1.5 + 0.9 + 0.35 = 4.25$

Thus, the mean number of rooms is 4.25 rooms. Once again, the mean does not have to be a whole number or a number which can actually occur in one house. It is, however, the expected value of the number of rooms, i.e. in some sense it is the centre of the probability distribution, and it means that, if we were to look at 20 houses, say, we would expect to find a total of around 85 rooms.

7.4.2 Variance of a Random Variable

As with the mean, the variance can be derived by taking a very large sample and evaluating the variance of the number of rooms in the sample. Alternatively, there is a formula which will give the variance without taking a sample.

Recall that one formula for the variance was of the form

$$\frac{\sum_{i=1}^{n} x_i^2 - \dfrac{\left(\sum_{i=1}^{n} x_i\right)^2}{n}}{n-1}$$

Well, once again, there is a computational form. If we let $V(X)$ denote the variance of the random variable X, then

$$V(X) = E(X^2) - (E(X))^2$$

$$\text{where } E(X^2) = \sum_{i=1}^{n} [x_i^2 P(X = x_i)]$$

In the example, $E(X^2) = 1^2 \times 0.05$
$$+ 2^2 \times 0.1$$
$$+ 3^2 \times 0.15$$
$$+ 4^2 \times 0.2$$
$$+ 5^2 \times 0.3$$
$$+ 6^2 \times 0.15$$
$$+ 7^2 \times 0.05$$

This equals $0.05 + 0.4 + 1.35 + 3.2 + 7.5 + 5.4 + 2.45$ which is 20.35 squared rooms. This is the expected value of X^2 and we have still to subtract the square of the mean.

Thus, the variance is $20.35 - (4.25)^2 = 20.35 - 18.0625$. Thus, we have a variance of 2.2875 squared rooms and, therefore, a standard deviation of 1.5124 rooms.

Example. A young boy in Sheffield has recently bought a new bicycle at a cost of £90. The police in the area reckon that, in a given year, the bicycle may be either stolen and lost, badly damaged, suffer a few minor bashes, or remain intact and unscathed. The police estimate the probabilities of each of these four possible outcomes as 0.2, 0.4, 0.3 and 0.1, respectively. Further, the losses in value attached to these outcomes (ignoring depreciation) are £90, £75, £30 and £0.

What is the expected annual loss on the bicycle? What is the variance of the annual loss? What is the maximum amount that might be worth paying for bicycle insurance.

If we let Y represent the loss (and we ignore depreciation), we have the following distribution.

y:	90	75	30	0
$P(Y = y)$:	0.2	0.4	0.3	0.1

The expected value of Y is equal to

$$90 \times 0.2 + 75 \times 0.4 + 30 \times 0.3 + 0 \times 0.1 = £57.$$

To get the variance, we need the expected value of Y^2 which equals $90^2 \times 0.2 + 75^2 \times 0.4 + 30^2 \times 0.3 + 0^2 \times 0.1$. This is equal to $1620 + 2259 + 270 + 0 = 4140$ squared pounds. Thus, the variance equals $4140 - (57)^2 = 4140 - 3249 = 891$ squared pounds. Thus, the standard deviation is £29.85.

As for insurance, if the expected annual loss is £57, it does not really make sense to pay more than this figure for insurance. Thus, in this case the expected value of the annual loss is the maximum figure to be paid in an annual insurance premium.

7.4.3 Chebyshev's Inequality

This section is a brief excursion into mathematics. However, it is very useful as the Soviet mathematician, Chebyshev, produced an inequality which represents, in effect, the worst that can occur.

Formally, it says that, for a random variable with a finite expected value and variance, the probability of an outcome being more than k standard deviations away from the expected value cannot be bigger than $1/k^2$.

(Note that it is possible to have infinite expected values and variances but you are unlikely to come across them.)

For example, let the random variable B denote the mid-month balance of ordinary deposit accounts in a large building society. Suppose that the mid-month balances have a mean of £4,800 and a standard deviation of £1,000. Let $k = 2$. We can immediately say that the probability of a mid-month balance being more than 2 standard deviations away from the mean is less than or equal to $1/2^2$. Thus, the probability of a mid-month balance being more than £2,000 (£1,000 $\times k$) away from £4,800 is no bigger than $\frac{1}{4}$. In other words, the probability of a mid-month balance being either less than £2,800 or in excess of £6,800 is less than or equal to 0.25 and, therefore, the probability of a mid-month balance lying between £2,800 and £6,800 is at least 0.75. Thus, we can immediately say that at least 75% of all accounts have a mid-month balance of between £2,800 and £6,800.

Now, we have previously stated that we usually get 95% of the observations within 2 standard deviations of the mean. However, what is being said here is completely different. This inequality is the "worst possible case". It says that, no matter what the distribution, we are guaranteed mathematically to have at least 75% of the observations between these two figures.

If we consider what happens when $k = 3$, we can follow a similar argument. The probability of a mid-month balance falling outside the interval £1,800 and £7,800 is less than or equal to $\frac{1}{9}$. Thus, we can say that there is a minimum of 88.889% of the balances in the interval £1,800 to £7,800.

Note that k does not need to be a whole number. For example, you may wish to know the minimum number of balances that will be found in the interval from £3,000 to £6,600. This corresponds to a value of k of 1.8. Thus, the maximum probability of being outside this interval is $1/(1.8)^2 = 0.309$. Thus, there will be a minimum of 69.1% of the balances within the interval £3,000 to £6,600.

7.4.4   Linear Functions of Random Variables

Often what is of interest is not the random variable but some function of it. Many types of functions may be of interest but linear functions, i.e. functions of the form $Y = a + bX$, are quite common and can be dealt with easily.

During the winter, a plumber in Glasgow works on a standby basis from his home. He is paid a retainer by his employers of £18 per day and is paid £15 for each time that he is called out to mend a burst pipe. He knows from past experience that the number of burst pipes for which he is called out has the following distribution. (Let X denote the number of "call-outs".)

x:	0	1	2	3	4
$P(X = x)$:	0.2	0.3	0.25	0.15	0.1

The plumber wishes to know the mean and standard deviation of his daily earnings.

If there are 0 call-outs, then he is paid £18. If there is 1 call-out, he is paid £33, and so on. Therefore, if we let Y denote the daily pay, Y has the following distribution.

y:	18	33	48	63	78
$P(Y = y)$:	0.2	0.3	0.25	0.15	0.1

We can evaluate the expected value and standard deviation of Y by using the formulas which we have discussed above.

Expected value of Y $3.6 + 9.9 + 12.0 + 9.45 + 7.8 = £42.75$

Expected value of Y^2 $64.8 + 326.7 + 576.0 + 595.35 + 608.4$
$$= 2172.25$$

Variance of Y $2171.25 - (42.75)^2 = 343.6875$

Standard deviation of Y £29.83

Now all this calculation was fairly simple but was made cumbersome by the size of the numbers. There is a simpler way to arrive at these answers using the relationship between X and Y.

As we saw, $Y = 18 + 15X$. In fact, it can be stated without proof that $E(Y) = 18 + 15E(X)$.

Also, $V(Y) = 15^2 V(X)$

The expected value and variance of X are far easier to calculate and, using the above relationships, should give the same answers.

Expected value of X $\quad 0 + 0.3 + 0.5 + 0.45 + 0.4 = £1.65$
Expected value of X^2 $\quad 0 + 0.3 + 1.0 + 1.35 + 1.6 = 4.25$
Variance of X $\quad 4.25 - (1.65)^2 = 1.5275$
Standard deviation of X $\quad £1.24$

Using these values, we get the expected value of Y to be $18 + 15(1.65) = £42.75$ and we get the variance of Y to be 225 times the variance of $X = 225 \times 1.5275 = 343.6875$.

This relationship may be utilized in considering the relationship between, for example, number of surveys and charges for surveys. All that is required is that we have a linear relationship between the two variables, X and Y, i.e. $Y = a + bX$ where a or b can, in some cases, be zero. Other applications where this is useful are in converting from imperial units to metric units, e.g. acres to hectares (and vice versa), and in converting from one currency to another currency, e.g. US dollars to pounds sterling.

Another place where this type of approach may help is in simplifying arithmetic. Suppose we had the following probability districution for X.

x:	990	1010	1020	1040
$P(X = x)$:	0.3	0.1	0.4	0.2

It might be far easier to work with $Y = -99 + 0.1X$ so that Y has the following distribution.

y:	0	2	3	5
$P(Y = y)$:	0.3	0.1	0.4	0.2

and then reconvert to X.

We can now state some general rules for working with linear functions of random variables.

In general, let $Y = a + bX$, where a and b are constant numbers and X and Y are random variables.

$$E(Y) = a + bE(X)$$
$$V(Y) = b^2 V(X)$$

Note that a and b need not be whole numbers; nor need they be greater than zero.

7.4.5 Sums and Differences of Random Variables

Suppose that the plumber in the above example is interested in his earnings for a five-day week. Let us further suppose that we can assume the five days to be independent of each other and that they have the same distribution as each other.

Let T be the random variable denoting the weekly total, and let Y_1 to Y_5 denote the five daily amounts.

$$E(T) = E(Y_1 + Y_2 + Y_3 + Y_4 + Y_5)$$

Now, without proof, it can be stated that the expected value of a sum is the sum of the individual expected values. In fact, this is true whether or not the random variables are independent.

Thus, $E(T) = E(Y_1) + E(Y_2) + E(Y_3) + E(Y_4) + E(Y_5)$

Each of the terms on the right hand side is equal to $E(Y)$ above. Thus, $E(T) = 5E(Y) = £213.75$

For the variance, we are trying to evaluate

$$V(T) = V(Y_1 + Y_2 + Y_3 + Y_4 + Y_5).$$

Now, the variance of a sum of random variables is equal to the sum of the variances, but only if the random variables are independent. We have assumed above that they are and so we can evaluate $V(T)$ to be five times $V(Y)$ which equals 10856.25. Thus, the standard deviation of his weekly earnings is £104.19.

Some readers may have thought that the variance of a sum of five identical random variables is 25 times the variance of one of the random variables. However, this is not so. There is a very important distinction to be drawn here between, at the simplest level, $2Y_1$ and $(Y_1 + Y_2)$. Recalling that the variance is a measure of deviation away from the mean, in the first case, any deviation will be multiplied by two and so be inflated while, in the second case, the deviation in Y_1 may be counter-acted by the deviation in Y_2.

Thus, in the case above, we were not multiplying the value of a random variable by five. Rather, we were looking at five independent random variables and adding up their values. By multiplying the variance by five instead of by twenty-five, we allow for the possibility, indeed the likelihood, of some deviations cancelling out others.

What does it mean to assume independence here? Well, suppose the plumber has four call-outs on Monday. Is the probability of four call-outs on Tuesday still 0.1; or might we think that perhaps we are in a cold spell and, therefore, there will be more burst pipes and so the probability of four call-outs on Tuesday is greater than 0.1?

In proceeding to consider differences between random variables, general rules can be stated for evaluating their mean and variance. For the mean, the expected value of a linear combination is the linear combination of the expected values. For example,

$$E(X + Y - W) = E(X) + E(Y) - E(W)$$

If we think of $+$ and $-$ as $+1$ and -1, then the above is a special case of

$$E(aX + bY + cW) = aE(X) + bE(Y) + cE(W)$$

which is valid for any constant numbers a, b and c.

Now, for the variance, the variance of a linear combination of independent random variables is the linear combination of the variances of the random variables, each one multiplied by the square of the number by which the random variable is multiplied. Thus, for the difference between two independent random variables,

$$V(X - Y) = V(X) + V(Y)$$

Note that we have a $+$ in the middle of the right hand side. There are three ways to try to understand why this is a $+$. First of all, if we recall that variances are squared quantities, then allowing the sign to be a minus sign would mean that either $V(X - Y)$ or $V(Y - X)$ would be a negative quantity; and negative variances are impossible. Secondly, one can realize that the variance is a measure of variability or unpredictability or uncertainty. Adding the variances represents the fact that there are now two sources of uncertainty. Thirdly, as can be seen immediately below, the more general formula has, as coefficient, the square of the original coefficient. Thus, if we think of $-$ as equal to a coefficient of -1, then we wish to have $(-1)^2 = +1$.

We can look at a more complicated linear combination such as

$$V(2X - 3Y) = 4V(X) + 9V(Y)$$

In general, we have

$$V(aX + bY + cW) = a^2V(X) + b^2V(Y) + c^2V(W)$$

7.5 Covariance and Correlation

Unfortunately, we can only look at the variance of a linear combination of random variables in such a simple manner if the random variables are independent. This is not always the case. Indeed, often in real data, the variables are so clearly dependent that one could not even justify an assumption of independence even if it were only to simplify the analysis.

When random variables are dependent and we wish to discuss their variance, we often need to involve another measure. This measure is the covariance.

Looking back through historical data, there is a strong association between the price of a bottle of rum and the salaries paid to government ministers. As one increases, the other tends to increase. We would say that these two variables are varying concurrently or that there is some covariation. We are not saying that one causes another—the subject of statistics is never capable of deducing cause and effect—but we would like to be able to quantify how close their covariation is. If it were very close then, it may be possible to accurately infer one from the other.

The basic measure of covariation is the covariance. For two random variables, X and Y, we define the covariation to be the product of the deviations of the two variables from their respective expected values.

Suppose that we are interested in the movement of property on an estate built just after the last war. We have the following discrete bivariate distribution for two variables X and Y where X is the number of rooms in a house and Y is the number of owners between 1970 and 1978.

		X			
		3	4	5	6
	1	0.11	0.08	0.06	0.05
Y	2	0.05	0.08	0.08	0.05
	3	0.05	0.09	0.05	0.25

If the covariation is the product of the deviations of the two variables from their respective means, we can write the covariations for one property as $(x_i - E(X))(y_j - E(Y))$. Further, we can define the covariance to be the average (mean) covariance over a series of trials. If we were to take a sample, we would evaluate the covariation for several resulting pairs (x_i, y_j), add them up and divide by the number of pairs.

However, once again, we need not actually collect a sample in order to evaluate the covariance. Given the probability distribution, we can evaluate the theoretical covariance in the same way as we did for the expected value. We can replace the frequency of occurrences divided by the total number of trials with a probability. As a result, it we denote the covariance of X and Y by $\text{Cov}(X, Y)$, we can evaluate it as

$$\text{Cov}(X, Y) = E(XY) - E(X)E(Y)$$

where $E(XY)$ is the expected value of a new random variable W, where $w_{ij} = x_i y_j$

Thus, in the above example, we can easily get the marginal distributions and so get $E(X) = 4.68$ and $E(Y) = 2.14$. For $E(XY)$, we must take each pair (x_i, y_j) and find W_y multiplied by $P(X = x_i$ and $y = y_j)$. Thus, for the first term, we have 3 times 1 times $0.11 = 0.33$, and so on until the last term of 6 times 3 times $0.25 = 4.5$. We then add these up to get 10.47.

The covariance is, therefore, equal to

$$10.47 - (4.68 \times 2.14) = 0.4548$$

Now, is this covariance large or small? Well, in reality, we cannot say for the covariance can take values from the very large negative numbers to the very large positive numbers. Thus, before we can interpret it, we must "cut it down to size" and standardize it. To do this we calculate the correlation.

7.5.1 Correlation

The main problem with the covariance is that it has no fixed scale. In fact, its scale is the same as that of the variables. To explain, suppose we have two variables X, Y measuring the dimensions of commercial sites and X, Y are measured in yards. We also have two variables V, W which measure the dimensions of the same commercial sites but in feet. Thus, $V = 3X$ and $W = 3Y$. If the covariance between X and Y is 1.8, say, then the covariance between V, W would be $3 \times 3 \times 1.8 = 16.2$. Thus, we need to standardize the covariance.

The correlation is, basically, a standardized version of the covariance. It is standardized so that the minimum value is -1 and the maximum value is $+1$.

A correlation of $+1$ is often called a perfect correlation or a perfect positive correlation. This would occur if, every time one variable increased by b units, the other variable increased by c units.

A correlation of -1 is often called a perfect negative correlation.

This would occur if, every time one variable increased by b units, the other variable decreased by c units.

In between these two is a range of values of the correlation which may occur in practice. In the middle of the range is the case of a zero correlation. Sometimes, when people see a correlation of zero, they assume that the two variables are independent. However, this is not necessarily so. If two variables are independent, they have zero correlation. The converse is not necessarily true for it is possible to have two variables which are dependent but have a zero correlation.

To actually calculate the correlation, we take the covariance and divide it by the square root of the product of the variances of the two variables.

$$\text{Thus, Corr}(X, Y) = E(XY) - E(X)E(Y)/\sqrt{V(X) \times V(Y)}$$

In the case above we can work out

$$
\begin{aligned}
E(X) &= 4.68 & E(X^2) &= 23.24 \\
E(Y) &= 2.14 & E(Y^2) &= 5.3 \\
V(X) &= 1.3376 & V(Y) &= 0.7204
\end{aligned}
$$

and the correlation equals 0.4548 divided by the square root of 1.3376 multiplied by 0.7204. This equals 0.4633. We are now able to provide some interpretation of this quantity for at least we know the bounds of this statistic.

Before leaving the subject of correlation for the present, we can present a different view of the coefficient of correlation. Suppose that we take a set of outcomes from repeated trials and plot these on a scatter plot. The coefficient of correlation will be close to $+1$ or -1 if these points lie very close to a straight line. In this way, we can think of the correlation coefficient as a measure of the degree of linear association between two sets of observations. On the other hand, it may be that some other relationship exists between the two variables and if this is not linear, then the correlation may not be able to detect it.

In the section which follows, we shall use the covariance for dealing with dependent random variables. However, later on, the correlation comes into its own and becomes far more important to and useful in the study of statistics.

7.6 Sums and Differences of Dependent Random Variables

Above, we discussed how to deal with the expected value and the variance of a linear combination of random variables. One problem which arose was that the formula quoted for the variance was applicable only if the random variables are independent of each other. This is generally not the case. Thus, we must introduce more general formulas.

$$V(X + Y) = V(X) + V(Y) + 2\text{Cov}(X, Y)$$
$$V(X - Y) = V(X) + V(Y) - 2\text{Cov}(X, Y)$$

These are general formulas in that they are valid whether the random variables are dependent or not. If the variables are independent, then the last term in each formula, the covariance, will be zero.

In general we have

$$V(aX + bY) = a^2 V(X) + b^2 V(Y) + 2ab\text{Cov}(X, Y)$$

One further point ought to be mentioned here while we are discussing covariance and correlation. It is possible to calculate the covariance and, therefore, the correlation from a set of data without actually constructing or knowing the relevant probability distribution.

For example, suppose that we had two sets of data which referred to the price a house is sold for (X) and the value put on it just prior to the property being put on the market (Y). Then, we could work out the covariance between these two sets of data by changing probabilities back into frequencies.

$$\text{Cov}(X, Y) = \frac{\Sigma x_i y_i}{n} - \frac{\Sigma x_i}{n} \frac{\Sigma y_i}{n}$$

and $\text{Corr}(X, Y) = \dfrac{\Sigma x_i y_i - (\Sigma x_i \, \Sigma y_i)/n}{\sqrt{\Sigma(x_i^2) - \dfrac{(\Sigma x_i)^2}{n}} \, \sqrt{\Sigma(y_i^2) - \dfrac{(\Sigma y_i)^2}{n}}}$

7.7 A Final Theoretical Word

Most of this chapter is fairly theoretical. However, no apology is offered. The theory is necessary to establish some formulas and relationships and to allow the reader to delve into problems of his/her own. Only by "getting one's hands dirty" can an intuitive feel be developed. Once this intuition is there, a lot of the theory can be set aside. This is either because the theory is obvious from an intuitive point of view—this happens to even the most reluctant student of statistics—or because the theory is so detailed and cumbersome that even those who choose to be statisticians by profession will refer to books for the details. As regards theory, the skill of a professional in any field is not the knowledge of it but the knowledge of where to find it and how to interpret it once it is found.

You will refer to this chapter on repeated occasions. It lays the foundations for most statistical theory and, therefore, will be of use not only later in this text but also for tackling problems and ideas which are encountered outside of this text.

Examples

1. The probability distribution of X, the number of passengers on a daily helicopter shuttle run from Abbotsinch airport at Glasgow to Gleneagles, is as follows:

x:	1	2	3	4
$P(X = x)$:	0.2	0.3	0.4	0.1

   (a) What is the probability that there are at least two passengers on the helicopter?
   (b) What is the probability that there are at least two passengers on the helicopter given that the helicopter has less than 4 passengers?
   (c) What is the mean and standard deviation of the number of passengers?
   (d) The fare for the trip is £28. Let Y denote the total fare revenue for a flight. What is the expected value of Y? What is the standard deviation of Y?
   (e) Suppose that cost to the company for each flight is £61. Let P represent the profit for a flight. What is the mean and standard deviation of P?
   (f) Let X_1 be the number of passengers on day 1 and X_2 be the number of passengers on day 2. Assume that the 2 days are independent, each with the probability distribution above. Construct the bivariate probability distribution.

2. Of a very large residential area, the bivariate distribution of the number of bedrooms, B, and the number of bathrooms (or toilets), T, per house is as follows:

				B		
		1	2	3	4	5
	1	0.19	0.445	0.22	0.001	0
T	2	0	0.015	0.06	0.058	0.006
	3	0	0	0	0.001	0.004

(a) Construct the marginal probability distribution for the number of bedrooms per house.

(b) Construct a distribution of conditional probabilities of the number of bedrooms in a house with only one bathroom.

(c) Calculate the correlation between B and T.

(d) Consider a new variable R, where R is the ratio of B to T i.e. it is the number of bedrooms "assigned" to each toilet. What values can R take? What is the probability distribution of R? What is the probability that R is bigger than 2?

(e) Consider a new variable D, where D is the difference between B and T i.e. $B-T$. Construct the probability distribution of D. What is the probability that D exceeds 1?

(f) What is the mean and the variance of D?

(g) Check these by using the mean and variance of R (and any other measure necessary) which you calculated in (c) and by using the respective formulas.

3. A Manchester restaurant accepts varying sizes of dinner parties. From past experience, the manager knows that N, the number of persons in a dinner party, has the following distribution:

n:	1	2	3	4	5	6	7
$P(N=n)$:	0.03	0.28	0.14	0.37	0.06	0.11	0.01

(a) What is the probability that a dinner party has an even number of customers?

(b) What is the probability that a dinner party has an even number of customers if we know that there are less than five people in it?

(c) Determine the mean and variance of N.

The restaurant serves customers a buffet dinner costing £7.25 per person plus a cover charge for the table of £2.50. Let C denote the total cost of the buffet dinner for a party.

(d) Express C in terms of N.

(e) What is the expected revenue to the restaurant from a dinner party?

(f) What is the standard deviation of C?

CHAPTER 8

Special Discrete Probability Distributions

In the previous chapter, we dealt with some discrete probability distributions which appeared to be fairly arbitrary; arbitrary in the sense that there is no pattern or rule for evaluating these probabilities—they are merely observed. In mathematical terms, we would say that there is no functional form to the probabilities.

Some probability distributions, however, are special and not arbitrary. In these probability distributions, the probability of the random variable taking a specific value fits into some pattern. These distributions can be considered special and if we can recognise either the pattern of probabilities or the situation where this pattern may occur, we can evaluate probabilities of events of interests in a quick and relatively simple manner.

Two of these probability distributions will be discussed in detail so that the reader can get some idea of how these special distributions may be used. After this, some mention of other special discrete distributions will be made.

8.1 Binomial Probability Distribution

Suppose that we carry out a series of trials (or tests or experiments) where the result of each one can be defined as being either a success or a failure. For example, we might consider a trial as a bid for a house or we might consider the taking of a professional exam as a trial. In these cases, success and failure should be obvious.

To return to the argument, suppose that we carry out a series of n trials where

   (i) n is known and fixed and does not depend on the results of any of the trials;
  (ii) The probability of success in each trial is constant; let us denote it p;
 (iii) the trials are independent of each other.

Let us suppose that, in a series of n trials, we have r successful trials and, therefore, $(n - r)$ trials which result in failure. The

Binomial probability distribution can very quickly give us the probability of this happening.

If the above conditions hold, then the probability of r successes out of n trials, each with probability p of success, is given by the Binomial formula and this probability is equal to

$$\binom{n}{r} p^r (1-p)^{n-r}$$

Thus, to calculate the probability, we must calculate three components and then multiply them together. The first component is the combinatorial term relating to the number of combinations of n items taken r at a time or, to put it another way, the number of ways of choosing r items from n items. In this case, we can consider it as the number of ways of choosing the r trials which are/were to result in a success from the total of n trials.

The second component is the probability of a success raised to the power r, the number of successes. This comes from the fact that, if the trials are independent, then the probability of a particular r being successes is p time $p \ldots r$ times i.e. p^r.

The third component is the probability of the failures. The probability of each failure is $1 - p$ as we can consider success and failure to be complementary events. This raised to the power $n - r$, which is the number of failures which actually occurred to give the third term of $(1 - p)^{n-r}$.

To get the actual probability of r successes out the n trials, we multiply the three components together.

Example. In November 1983, a building contractor submitted tenders for 6 pieces of work. On past evidence, he reckons that he has a 0.3 chance of being awarded each contract and, as the contracts are in six different parts of the country, he assumes that they are independent. What is the probability that he awarded all 6 contracts? What is the probability that he is awarded exactly 3 of them? If he requires at least 3 to cover his running costs for the month, what is the probability that he fails to meet running costs for November?

Firstly, before we perform any calculations, we must satisfy ourselves whether the Binomial probability distribution is applicable. We certainly have a series of trials where the result of each can be considered as either a success or a failure. Further, we are told to assume that they are independent and that the probability of a success in each one is 0.3. (In reality, it would be strongly advisable to make some effort to verify or substantiate these assumptions). We must also assume that n is fixed in the sense that, whatever the outcomes of the trials, there were 6 trials. This would not be the case if, for example, failure to be awarded the first four contracts resulted in the company becoming insolvent and ceasing to be in operation.

We can now proceed to the actual calculation of the desired probabilities. For the first part of the question, it is merely a matter of putting in values for n, p and r in the formula.

$$n = 6 \qquad p = 0.3 \qquad r = 6$$

(we are interested in the probability of winning all the contracts)

$$\binom{6}{6} = 1, (0.3)^6 = 0.000729, \text{ and } 0.7^{(6-6)} = 0.7^0 = 1$$

Thus, the probability that we wish is equal to 0.000729.

The second part of the problem asked for the probability of exactly 3 successes. Thus, we change $r = 6$ to $r = 3$.

$$\binom{6}{3} = 20, 0.3^3 = 0.027, \text{ and } 0.7^{(6-3)} = 0.7^3 = 0.343$$

Thus, the probability that we wish here is equal to $20 \times 0.027 \times 0.343 = 0.18522$.

In the last part of the question, we are asked for the probability that he fails to meet running costs for the month, i.e. that he does not get 3 or more of the contracts in that month. Thus, we wish the probability of 2 or fewer of the contracts which is equal to the probability of getting 0 of them plus the probability of getting 1 of them plus the probability of getting 2 of them.

$$P(R = 2) = 0.324135$$
$$P(R = 1) = 0.302526$$
$$P(R = 0) = 0.117649$$

and, therefore, the probability of failing to meet running costs for the month of November 1983 is the sum of these three probabilities which equal 0.74431. (Check that your answer is the same).

Having looked at an example, we can examine whether the assumptions which are necessary are realistic and what their consequences are. Firstly, we have the assumption that n is fixed beforehand. This is fair enough in most cases but, if we consider the case of the number of trials being the number of clients which an estate agent takes to see a property before it is sold, then the number of trials, i.e. the number of clients who view, is not fixed. In fact, here the number of trials depends directly upon the result of the trials. (This would be a situation where the Geometric Distribution is applicable—see later).

The second assumption concerns the constancy of the probability of success in each trial. In this case, there has to be no learning effect. Learning effects are common in sitting professional exams, for example. Also, the contractor in the above example may feel that the probability is not constant if one awarding authority is in charge of all of the contracts. The authority may feel that it is advisable to share out the work and so, if the contractor gets the first contract, there is a decreased probability of getting successive contracts. Similarly, if he does not get the first one, then he is slightly more likely to get one of the other five.

This leads on to the third assumption; for the above is clearly a case of statistical independence. For the Binomial to be applicable, we must have statistical independence. The result of any of the trials cannot affect the result of any other of the trials. This would not be the case if, for example, the awarding authority stipulated that no contractor should be awarded more than three of the contracts—here, if the first three were successful, the probability of getting the fourth is no longer 0.3 but is 0.0 as it has been deemed impossible. Thus, the probability has been changed by the results of the other trials—i.e. we have non-constant probability and statistical dependence. On the other hand, if the awarding authority stated that, to cut down on administrative costs, one contractor will be awarded all the contracts then, as there is a 0.3 probability of being awarded the first contract, there is also a 0.3 probability of being awarded the fourth or the sixth contract. However, they are clearly not independent. There is, in fact, a perfect correlation here as either the bids are all successful or they are all unsuccessful.

8.2   Mean and Variance of a Binomial Variable

It will be of interest to know what is the mean and variance of the number of contracts which he is awarded. This will be especially so if the contractor wants to know the mean and variance of the number of contracts that he is awarded in a whole year or a five-year planning cycle.

To evaluate the mean or the expected value, we could work out the probabilities of the seven possible outcomes (0–6 contracts), and then use the formula for the expected value given in the previous chapter. However, in the case of a Binomial distribution, this is not necessary as there is a far simpler form. Roughly speaking, if we submit tenders, each of which has a 0.3 chance of succcess, and we do this 6 times, then we ought to expect about 6×0.3 successful bids = 1.8. In fact, this is the correct answer.

Mean of a Binomial variable with n trials, each with a probability of success of p
$$= np$$

Remember that this is a mean or an expected value and need not be a whole number. What it says is that the distribution is situated around 1.8. Thus, in 10 similar months, he would expect to be awarded 18 contracts. (Note, that if we had rounded up the 1.8 to 2, we would now get an overestimate for the figure for ten months of 20 instead of 18). Of course, this is only a mean figure or an expected figure and does not relate how far away from this that we might be. To obtain some idea of the spread, we look at the variance.

Once again, we could work out the variance from the seven probabilities and the formula in the previous chapter. However, the Binomial distribution affords us another quick way to do the calculation.

Variance of a Binomial variable with n trials, each with a probability of success of p
$$= np(1 - p)$$

In the above example, the variance is $6 \times 0.3 \times 0.7 = 1.26$. Thus, the standard deviation is 1.1225 and we can now get some idea of the spread of the number of contracts. For a month, the number of contracts awarded to the contractor has a mean of 1.8 and a standard deviation of 1.1225. For the ten month period, the number of contracts awarded to the contractor has a mean of 1.8 and a standard deviation of 3.5496. (Check that you can get this answer— we assume that the ten months are independent and so the variance for the total of ten months is 12.6.)

Example. The same contractor is also submitting tenders in another area of the country. He is submitting for 10 contracts but estimates that he has only a 0.2 probability of getting each one. He is interested in the total number of contracts that he is awarded from the 16 tenders.

If we assume that the two areas and their awards are independent of each other, then we can make some progress on this problem. The expected total number of contracts awarded is the sum of the expected values for the two areas. (We do not need the assumption of independence here.) Thus, the expected total number is $1.8 + 2.0 = 3.8$. To calculate the variance, we use the assumption of independence and merely add the variances of the two components. Accordingly, the variance of the total number of contracts awarded is $1.26 + 1.6 = 2.86$. Therefore, the standard deviation is 1.6912 contracts.

However, if, for example, we wish to know the probability of being awarded a total of 3 contracts, we cannot do it as simply as we might expect. The total number does not have a Binomial distribution. We do have a series of independent trials but the probability of success in each one is not constant. Some have a probability of success of 0.3 and some have a probability of 0.2. Thus, to calculate the probability of being awarded 3 contracts, we must evaluate the probability of each of the four ways of doing this, and, using the mutual exclusiveness, add them up

i.e. P(3 out of the 6 and 0 out of the 10)

$+ P$(2 out of the 6 and 1 out of the 10)

P(1 out of the 6 and 2 out of the 10)

$+ P$(0 out of the 6 and 3 out of the 10)

Considering the first of the above terms, this equals P(3 out of the 6) times P(0 out of the 10) and each of these two probabilities can itself be evaluated from a Binomial distribution. The probability of exactly 3 contracts can be evaluated to be

$$0.019893 + 0.086998 + 0.091360 + 0.023683 = 0.221934$$

8.3 Poisson Probability Distribution

Another special distribution for discrete random variables is the Poisson distribution, which is named after a French mathematician. Unlike the Binomial distribution which has specific constraints put upon its applicability, the Poisson is applicable, in general, in situations where events happen at random either in time or in space. Thus, we might consider that the breakdowns in a piece of office equipment occur at random in time. We also might have flaws in a pipeline or in a piece of cloth occurring at random along the length. Further, we might consider that currency devaluations occur at random in time. All of these may be modelled by a Poisson distribution i.e. the related variable has a Poisson distribution.

Let us consider the case of a Planning office where applications for planning permission for a district in the south of London arrive at random in time. We can consider the random variable, X, to denote the number of applications which arrive in a given interval of time, and that X has a Poisson distribution with a parameter m. (A parameter is a quantity which is fixed for a specific example and the value of which is substituted into a formula when necessary. An example of a parameter is a rate of exchange which can be substituted into a currency conversion calculation when required. The value of this parameter may change from one trading day to the next.)

Let us suppose that we wish to know the probability of there being 4 applications in an interval of time. As in the case of the Binomial distribution, the Poisson distribution also has a formula for evaluating probabilities.

$$P(X = x) = \frac{e^{-m}m^x}{x!}$$

(Recall the meanings of e and ! from chapter 4.)

Thus, in the case of $X = 4$, once we know m, we can evaluate a probability. Suppose that m is 5. Then, the probability of getting 4 applications is $(e^{-5}5^4)$ divided by 4! This equals 0.00674 times 625 divided by 24. The resulting probability is 0.1755 approximately. Similarly, the probability of exactly 6 applications is 0.00674 times 15,625 divided by 720 which equals 0.1462 approximately.

Before looking at a full example, an obvious question is "what is this parameter m?" In fact, this parameter is the mean or the expected value of the random variable. Thus, in the above example, we expect, on average, five applications. Once again, we could work out the mean from first principles by working out the probabilities of all the possible numbers of applications. However, this is rather difficult to do as the number is infinite and so we would have to use some quite powerful mathematics to evaluate the mean. We can note here the most basic difference between the Binomial distribution and the Poisson distribution. That is that the former has a finite number of outcomes of the random variable $- n + 1$ where there are n trials—while the Poisson distribution can cope with random variables with an infinite number of possible outcomes.

Thus, in most cases of using the Poisson distribution, one will have to be given the mean or will have to estimate it from other sources—resorting to infinite mathematics in the real world is not usually practical.

If we let X denote a random variable which has a Poisson distribution, then

Variance of X = Expected value of X

This is a property which is peculiar to the Poisson distribution— that the mean and the variance are the same. Actually, peculiar is the wrong word as a discrete distribution can be easily constructed to have this property. However, none of the common special discrete distributions have this property.

Having looked at the interpretation of the parameter, we can now look more closely at an example.

Example. The number of telephone enquiries received, during office hours, by a Brighton estate agent, has a Poisson distribution with a mean of 1.3 per 5 minutes. What is the probability that, in a selected 5-minute period, there are

  (i)   3 telephone enquiries
 (ii)   2 telephone enquiries
(iii)   0 telephone enquiries

Looking at the first part, we have a Poisson distribution with a mean of 1.3. Thus, from the formula, we can calculate the probability. $e^{-1.3}$ is 0.2725 and for the three parts of the question, we calculate $(1.3)^x$ *and* $x!$ where x takes the values 3, 2 and 0.

  (i)   0.2725 times 2.197 divided by 6 = 0.100 (to 3 decimal places)
 (ii)   0.2725 times 1.69 divided by 2 = 0.230
(iii)   0.2725 times 1 divided by 1 = 0.2725

If we consider the number of telephone enquiries in a half hour, what is the probability of there being

 (iv)   9 telephone enquires
  (v)   6 telephone enquires
 (vi)   0 telephone enquires

There is still a Poisson distribution here but the mean has changed. If we expect 1.3 per 5-minute period, we expect 7.8 per half hour. Thus, the answer to part (iv) is

$$\frac{e^{-7.8}(7.8)^9}{9!} = 0.121$$

The answers to the other two parts are 0.128 and 0.00041.

This last answer could be evaluated in another way which sheds light on how probabilities and events are linked. Part (vi) asked for the probability of no enquiries in a half hour period. This could be interpreted as being the probability of no enquiries in six successive (and independent) 5-minute periods. This is, using part (iii), equal to $(0.2725)^6 = 0.00041$.

8.4 Cumulative Probabilities

One further point, in the problem above, is that usually we are not interested in the probability of exactly one event occurring. For example, in the above problem, the estate agent may be interested in whether or not the staff on the telephone switchboard can handle the calls. If the estate agent reckons that a maximum of 11 calls can be handled per half hour, the question of interest is the probability of 12 or more enquiries in a half hour for this could involve an enquirer receiving an "engaged" signal or not being attended to properly. To evaluate this probability there are at least three approaches.

The first approach is to evaluate the probability of exactly 12 enquiries and the probability of exactly 13 enquiries and so on and add them all up. However, this runs into a problem as a Poisson variable has, in theory, an infinite maximum value.

The second approach is to evaluate the probability of the complementary event "11 or fewer enquiries" and subtract the probability from 1.

The third approach is to utilize one of the many sets of tables of these probabilities which are available. These tables give, for certain values of m, the probability of "x or more events" or of "x or less events". Of course, if the value of m we wish is not in the tables, then we must resort to one of the two approaches above.

These tables also contain cumulative probabilities for the Binomial distribution for various values of n and p. Better sets of tables are those by Neave and by Murdoch and Barnes.

8.5 Other Special Discrete Probability Distributions

The Binomial and Poisson distributions were discussed in depth so that an idea of the use and application of these distributions might be gained. Many other special discrete distributions exist. While we shall not discuss them here, it will be useful to mention a few together with the type of situation which might be modelled by such a distribution.

The distributions below are not discussed in depth because most of them (and others) can be found in any good book on Statistics and Probability or on Applied Probability or on Probability Models. Tables of probabilities and cumulative probabilities are also available for some of these distributions while, for others, the mathematical form is of a particular simple nature.

8.5.1 Geometric Distribution

Here, we are carrying out a series of independent trials until a success occurs. The random variable is the total number of trials necessary.

8.5.2 Negative Binomial Distribution

Here, we are carrying out a series of independent trials until the kth success occurs. (If $k = 1$, this reduces to a geometric distribution.) The random variable is the total number of trials required to record k successes.

8.5.3 Multinomial Distribution

Here, we are carrying out a series of independent trials each of which has k possible outcomes. (If $k = 2$, this reduces to a Binomial distribution.) The random variable is, in fact, multidimensional in that we can evaluate the probability of n_1, outcomes of type 1, n_2 outcomes of type 2, . . . and n_k outcomes of type k.

8.5.4 Hypergeometric Distribution

This is similar to the Binomial distribution except that the probability changes from trial to trial. This takes account of the fact that, if we are dealing with a small population, the probabilities will change from trial to trial. For example, suppose we are considering twenty villas and thirty bungalows. If we choose five houses, the probability of the first one being a villa is 0.4 but the probability of the second one being a villa is $19/50$ (0.38) or $20/49$ (0.408) depending on whether the first one was a villa or not. Normally, when we apply the Binomial distribution to such a case, we assume that the population is so large that these differences are negligible.

8.5.5 Discrete Uniform Distribution

This applies when we have a trial whose possible outcomes are equally likely.

8.6 A Final Word on Special Discrete Distributions

While the above-mentioned special discrete distributions call for some extra work in evaluating probabilities, they are also of great use. The existence of a precise mathematical form means that several of the properties in which we may be interested can be calculated quite easily. For example, some of the probabilities are quite simple to calculate and, in some of the cases, the mean and variance can be evaluated from a simple formula.

Secondly, although extra effort may be required from time to time, the use of tables as mentioned in 8.4 and 8.5 will greatly help.

Thirdly, the use of these special distributions, either exactly or as an approximation to a discrete distribution, is becoming more and more widespread. In particular, several legal practices, some surveying practices, and many accountancy firms are beginning to use distributions such as these to provide models of their operations. These models are being found to be of great use in such diverse areas as manpower planning, budgetary control, the purchasing, maintenance and reliability of equipment, forecasting, and general corporate decision-making. The range of applications is very large; the only problem is gaining the knowledge and understanding which is necessary to use and interpret probability, in general, and probability models, in particular.

Hopefully, this text will go some way to providing you with that knowledge and understanding, thus giving you a good start on the road to technological management along which we are being directed every day.

Problems

1. A Transylvania Law Society report stated recently that 30% of all complaints were about high conveyancing charges. In a typical month, nine complaints were received. Determine the probability that exactly four of these complaints were about high conveyancing charges.
2. One of the largest building societies has announced that, of new mortgages, 15% are for properties in Scotland. In the last period under study, 50 clients were given mortgages. What is the probability that exactly 8 were for properties in Scotland? By examining a set of tables, determine the probability that more than 11 were for Scottish properties. What is the expected number and standard deviation of the number of properties in Scotland?
3. In 8.1, three conditions are stated which must hold for a Binomial probability distribution to be valid. For each condition, think of a situation where this condition does not hold but the other two do. In each of these three situations, what might be done to evaluate probabilities of interest?

4. Local authority surveyors are concerned about the high number of individuals who, having built their own home, have their first application for a completion certificate refused. Indeed, it is estimated that about 22% of these refusals are for failures in external (site) works such as drains and manholes, paths and paved areas, and fences, boundary walls, and gates. Last month in one district, there were seven refusals. What is the probability that fewer than three of these refusals were for failures in external works?

5. Enquiries about a particular property are reckoned to be made at random so that the number of enquiries in a day has a Poisson distribution with a mean of 2.1. What is the probability that, in a day, there are
    (i) no enquiries?
    (ii) four enquiries?
   What is the probability that, in a five-day week, there are
    (iii) two enquiries?
    (iv) ten enquiries?

6. Breakdowns in a computer system in a legal practice are known to occur at random with a mean number of breakdowns per month of 0.9. What is the probability that, in a given month, there are
    (i) no breakdowns?
    (ii) one breakdown?
    (iii) two or fewer breakdowns?
   What is the probability that, in a year, there are ten breakdowns?

CHAPTER 9

Continuous Random Variables

9.1 Introduction

In 7.1, we discussed random variables, their definition and the distinction between being discrete and continuous. Chapter 7 and 8 went on to deal in depth with general discrete probability distributions and particular discrete probability distributions. This chapter deals with continuous probability distributions in general and the three particular continuous distributions which we shall be using frequently in subsequent chapters. A warning is necessary here. Some of this chapter requires a knowledge of calculus. However, the three distributions can be understood without calculus provided sections 9.2–9.4 are at least read.

9.2 Continuous Probability Distributions

A continuous random variable has an infinite number of possible values as it can take any value along a continuous scale. As a result, if we were to have $P(X = x)$ greater than zero for an infinite number of values, the sum would exceed 1. Thus, we must say that $P(X = x) = 0$ for all values of x.

Instead of representing probabilities as we do in a discrete way as in Figure 7.1, we construct a continuous function to represent probability.

As an example, we can consider a random variable X which represents the amount of snow which falls on a week in January. Suppose that we know that the maximum snowfall is $3\,m$. We can represent the probability by a mathematical function $f(x)$.

To ensure that this can represent probability, we need to impose two constraints: that this function never has a negative value and that the total area under the curve is 1.

Once these constraints are imposed, we can answer questions about the random variable by examining the relevant areas. For example, if we wished to know $P(X \geqslant 2.1)$, we need to measure the area under the curve between 2.1 and 3.

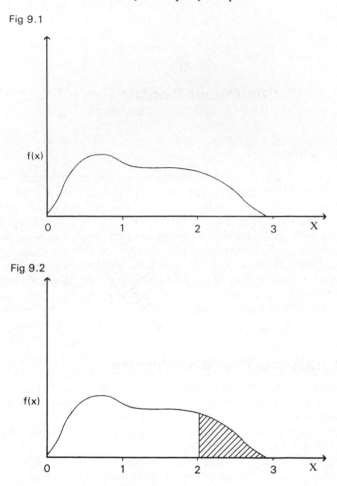

Fig 9.1

Fig 9.2

Similarly, $P(0.8 \leqslant X \geqslant 1.7)$ is the area under the curve between the two points.

Note that as $P(X = x) = 0$ for all values of x, there is no difference between $P(X \geqslant 2.1)$ and $P(X > 2.1)$ if X is a continuous random variable.

Definition: $f(x)$ described above is commonly referred to as the probability density function (p.d.f.) of a continuous random variable X and is a mathematical function subject to

$$f(x) \geqslant 0 \quad \text{for all values of } x$$

and $\qquad\qquad \int f(x) = 1$

where $\int$ is the mathematical sign for integration, i.e. for summing an area.

9.3 Cumulative Probability Distributions

As in the discrete case, we are often interested in $P(X \leqslant x)$.

This is often called the cumulative distribution function (c.d.f.) and is denoted $F(x)$ which can also be evaluated by integration as

$$\int_{-\infty}^{u} f(u(\mathrm{d}u$$

Recall that this function must be $\geqslant 0$ for all values of x and, in fact, for values less than 0 and greater than 3, $f(x) = 0$.

9.4 Properties and Functions of a Continuous Random Variable

Roughly speaking, we can treat integration in the continuous case as being analogous to summation in the discrete case. As a result, many similar functions can be investigated.

9.4.1 Mean of a Continuous Random Variable

"What is the mean snowfall?" This can be easily evaluated by

$$E(x) = \int_{-\infty}^{\infty} xf(x)\mathrm{d}x$$

(This is analogous to $\Sigma xp(x)\mathrm{d}x$)

9.4.2 Variance of a Continuous Random Variable

Once again, it is often easier to use the fact that

$$V(x) = E(x^2) - [E(x)]^2$$

$$E(x^2) = \int_{-\infty}^{\infty} x^2 f(x)\mathrm{d}x$$

and we have $E(x)$ from above.

9.4.3 An Example of a P.D.F.

As an example of a p.d.f. for the snowfall, we can consider

$$f(x) = \frac{2(x+1)}{15} \quad 0 \leqslant x \leqslant 3$$

$$0 \quad \text{otherwise}$$

Fig 9.3

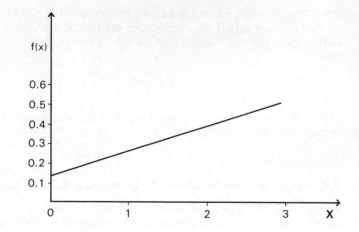

Before doing any analysis, we must check that the two constraints are met. Clearly all values in the range 0 to 3 are positive and all others are zero so it looks as if the first constraint is met. As for the second, we need to integrate,

$$\int_{-\infty}^{\infty} f(x)\,dx = \int_0^3 \frac{2(x+1)}{15}\,dx$$

$$= \frac{2}{15} \int_0^3 (x+1)\,dx = \frac{2}{15}\left[\frac{x^2}{2} + x\right]_{x=0}^{x=3}$$

$$= \frac{2}{15}\left(\frac{9}{2} + 3\right) = 1$$

We can now answer some questions. What is $P(X \geqslant 2.1)$? This equals

$$\int_{2.1}^3 \frac{2(x+1)}{15}\,dx = \frac{2}{15}\left[\frac{x^2}{2} + x\right]_{x=2.1}^{x=3}$$

$$= \frac{2}{15}\left(\frac{9}{2} + 3 - \frac{4.41}{2} - 2.1\right) = 0.426$$

What is $P(0.8 \leqslant X \, 1.7)$? This equals

$$\frac{2}{15}\left[\frac{x^2}{2} + x\right]_{x=0.8}^{x=1.7}$$

$$= \frac{2}{15}\left(\frac{2.89}{2} + 1.7 - \frac{0.64}{2} - 0.8\right)$$

$$= 0.27$$

What is the median value of X? i.e. $P(X \leqslant \text{median}) = 0.5$?

$$\int_0^{\text{median}} \frac{2(x + 1)}{15} \, dx = 0.5 \Rightarrow \frac{2}{15}\left(\frac{x^2}{2} + x\right)_0^{\text{median}} = 0.5$$

$$\Rightarrow \frac{2}{15}\frac{\text{median}^2}{2} + \frac{2}{15}\text{median} = 0.5 \Leftrightarrow \text{median}^2 + 2\,\text{median} = 7.5$$

$$\Leftrightarrow m^2 + 2m - 7.5 = 0 \Leftrightarrow m = 1.915$$

i.e. by solving an equation with the median as the unknown variable and then solving a quadratic equation, the median turns out to be 1.915. In other words, 50% of the time the snowfall will exceed 1.915 m.

What is $E(X)$? This is

$$\int_0^3 x\frac{2(x + 1)}{15} \, dx = \int_0^3 \frac{2}{15}(x^2 + x)dx$$

$$= \frac{2}{15}\left[\frac{x^3}{3} + \frac{x^2}{2}\right]_{x=0}^{x=3} = \frac{2}{15}\left(9 + \frac{9}{2}\right) = \frac{13}{15}$$

What is $V(X)$? First we need

$$E(X^2) = \int_0^3 x^2\frac{2(x + 1)}{15} \, dx$$

$$= \frac{2}{15}\left[\frac{x^4}{4} - \frac{x^3}{3}\right]_{x=0}^{x=3} = 3.9$$

Thus, the variance of X is $3.9 - (13/15)^2 = 3.1489$ and the standard deviation is 1.7745 m.

We shall now go on to discuss three particular distributions: the Normal distribution, the X^2 distribution and the F distribution.

These will not, in general, require calculus or any of the manipulations as above. This is all avoided as these three distributions have tables of probabilities included in most statistics books and this one is no exception.

9.5 Normal Probability Distribution

The Normal probability distribution is almost certainly the most important distribution in Statistics. Its importance is due not only to some very important and far-reaching theory but also to its frequent occurrence in practice.

The Normal distribution cannot simply be represented by one curve. Rather it is an infinite number of curves, often called a family of curves, all of which have the same basic shape but differ in their location and/or their spread. The basic shape of the Normal distribution is as in Figure 9.4a but in Figure 9.4b we see three

Fig 9.4a

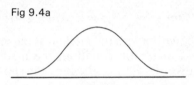

Normal curves. Curves A and B have the same spread but differ in location. Curves A and C have the same location but differ in spread. We can talk of location in a fairly relaxed way as, in a Normal distribution, the mean, median and mode (the point at which the peak occurs) have the same value.

Fig 9.4b

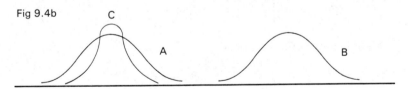

The p.d.f. of a Normal distribution is

$$f(x) = \frac{1}{\sqrt{2\pi\sigma^2}} e^{-\frac{(x-\mu)^2}{2\sigma^2}}$$

and, in fact, each different Normal curve is identified by two parameters, μ and σ^2, i.e. the mean and variance.

Once we have these we can draw the curve as in Figure 9.5,

Fig 9.5

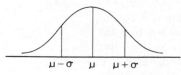

i.e. the points on the curve where it starts to turn up instead of down are one standard deviation from the mean.

9.5.1 Standard Normal Probability Distribution

We shall now look at a particular Normal distribution and we shall look at the simplest one possible. This has a mean of zero and a variance of 1 and is usually called the standard Normal distribution.

We usually denote a variable which has a standard Normal distribution by the letter Z and we can now try to find $P(Z > 1)$, $P(Z > 1.78)$, $P(Z > 0.31)$, $P(-2.1 > Z > 0.35)$, etc.

In table 1 at the end of this text, values of $P(Z > a)$ are given for a ≥ 0. We can use this table, together with the idea of complementary events and symmetry, to be able to evaluate many probabilities of interest. In fact, seven types of problems occur commonly and we can easily develop rules for solving them.

Rule 1: $P(Z > a)$ for $a \geq 0$

These probabilities are the entries in the table. Check that

Fig 9.6

$P(Z > 1) = 0.1587,$
$P(Z > 1.78) = 0.0375,$
$P(Z > 2.16) = 0.01539$

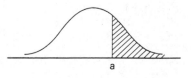

Rule 2: $P(Z < b)$ for $b \geq 0$

Here, we use the idea of complementary events and say that $P(Z < b) = 1 - P(Z > b)$. (Recall that we are taking $P(Z = b)$ to be zero.) Check that

Fig 9.7

$P(Z < 1.4) = 0.9192,$
$P(Z < 0.29) = 0.6141,$
$P(Z < 2.05) = 0.9798$

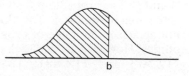

Rule 3: $P(Z < c)$ for $c < 0$

Here, we use the symmetry of the Normal distribution. Looking at figure 9.9, we can see that the area to the left of c is the same as the area to the right of $-c$, which is, of course, positive.

Thus, we can say that, for $c < 0$, $P(Z < c)$ is equal to $P(Z > -c)$ which we can evaluate using Rule 1. For example, $P(Z < -0.63)$ is equal to $P(Z > 0.63)$ which is 0.2643. Check that

$$P(Z < -1.88) = 0.0301,$$
$$P(Z < -0.61) = 0.2709$$

Rule 4: $P(Z > d)$ for $d < 0$

Here, we use the two ideas together. Firstly, we can say that $P(Z > d)$ is equal to $1 - P(Z < d)$. Secondly, if $d < 0$, we have seen in Rule 3 that $P(Z < d) = P(Z > -d)$. Thus, $P(Z > d)$ is equal to $1 - P(Z > -d)$.

For example, $P(Z > -0.21) = 1 - P(Z < -0.21) = 1 - 0.4168 = 0.5832$. Check that

$$P(Z > -1.46) = 0.9279,$$
$$P(Z > -0.59) = 0.7224,$$
$$P(Z > -1.03) = 0.8485$$

Rule 5: $P(e < Z < f)$ for $e > 0$, $f > 0$

We now have to introduce another simple idea. In figure 9.13, we have two regions marked A and B. The probability we wish is the area of region A. This does not appear in the table of probabilities. However, we can look up $P(Z > e)$ and $P(Z > f)$. These

Fig 9.8

Fig 9.9

Fig 9.10

Fig 9.11

Fig 9.12

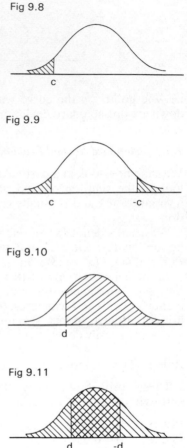

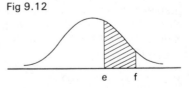

correspond, respectively, to the
areas of regions A and B
together, and to the area of
region B alone. The difference,
therefore, is the area of region A.

Thus, $P(e > Z > f)$ is equal to
$P(Z > e) - P(Z > f)$. For example, $P(0.6 > Z > 1.4)$ is equal to
$P(Z > 0.6) - P(Z > 1.4)$ which
equals $0.2743 - 0.0808$ which
equals 0.1935. Check that

$P(0.21 > Z > 0.93) = 0.2406,$
$P(0.59 > Z > 1.92) = 0.2502.$

Fig 9.13

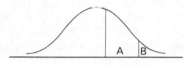

Rule 6: $P(g < Z < h)$ for $g < 0$,
$h < 0$

If, once again, we use the idea
of symmetry, we can see that
$P(g < Z < h)$ is equal to
$P(-h < Z < -g)$.

Thus, we can use Rule 5 which
states that this probability equals
$P(Z > -h) - P(Z > -g)$. For
example, $P(-0.73 < Z < -0.16)$
$=$ $P(0.16 < Z < 0.73)$. This
equals 0.2037. Check that

$P(-1.58 < Z < -1.14) = 0.07,$
$P(-1.36 < Z < -0.54) = 0.2077.$

Fig 9.14

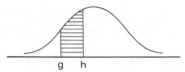

Fig 9.15

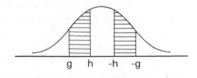

Rule 7: $P(i < Z < j)$ for $i < 0$,
$j \geqslant 0$

Here, we use the idea developed in Rule 5 and rewrite
$P(i < Z < j)$ as $P(Z > i)$
$- P(Z > j)$. However, i is negative and so we use Rule 4 and
reexpress $P(i < Z < j)$ as
$1 - P(Z > -i) - P(Z > j)$. For
example, $P(-0.56 < Z < 0.2)$
$= 1 - P(Z > 0.56) - P(Z > 0.2)$.
This equals $1 - 0.2877 - 0.4207$
which is equal to 0.2916. Another
way of thinking about this rule is

Fig 9.16

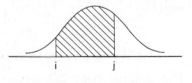

to use the fact that the total area under the curve is 1 and to subtract the area of regions which do not fall between i and j, i.e. $1 - P(Z < i) - P(Z > j)$ and to use Rule 3 to reexpress $P(Z < i)$ as $P(Z > -i)$. Check that $P(-0.36 < Z < 2.04) = 0.6199$, $P(-0.7 < Z < 0.4) = 0.4134$.

To recap, the seven rules which we have developed are as follows. Let $T(a)$ be the tabulated value at a.

1. $P(Z > a)$ $a \geqslant 0$ $= T(a)$
2. $P(Z < b)$ $b \geqslant 0$ $= 1 - T(b)$
3. $P(Z < c)$ $c < 0$ $= T(-c)$
4. $P(Z > d)$ $d < 0$ $= 1 - T(-d)$
5. $P(e < Z < f)$ $e \geqslant 0, f \geqslant 0$ $= T(e) - T(f)$
6. $P(g < Z < h)$ $g < 0, h < 0$ $= T(-h) - T(-g)$
7. $P(i < Z < j)$ $i < 0, j \geqslant 0$ $= 1 - T(-i) - T(j)$

We can also change the question around and ask what value is exceeded 15% of the time. In other words, we wish to find k such that $P(Z > k) = 0.15$. Table 1 contains such probabilities and so we have to search around in the heart of the table until we find 0.15 or something very close to it. The closest we come is 0.1492 at $k = 1.04$.

However, in asking a question of this type, we may have to use one of the above rules to reexpress the question in a suitable form. For example, we may wish to find k such that $P(Z > k) = 0.64$ or to find k such that $P(-k < Z < k) \, 0.83$.

In the first case, if $P(Z > k) = 0.64$, then k is clearly negative and so we know that $P(Z > k) = 1 - P(Z > -k)$ which equals 0.64. Thus, $P(Z > -k) = 0.36$ and so $-k = 0.36$. Thus, $k = -0.36$. In the second case, we have that $1 - P(Z > k) = 0.83$. Thus, $2P(Z > k)$ must equal 0.17 and so $P(Z > k) = 0.085$. k, therefore, is close to 1.37.

9.5.2   *A Warning About Tables of Probabilities*

The table on page 263 gives values of $P(Z > a)$ for $a \geqslant 0$. This is probably the most common type of table for the standard Normal probability distribution. However, other types of tables which occur frequently give $P(Z < a)$ or $P(0 > Z > a)$. If you use another set of tables, make sure that you know exactly what the entries correspond to. You should be able to use the other sets of tables, with caution, but the seven rules we developed will have different, but analogous, versions.

9.5.3 *Non-Standard Normal Distributions*

1.—Mean not equal to zero.

Suppose that we have a Normal random variable, X, with a mean of 5 and a variance of 1. What is the probability that this variable takes a value in excess of 6? Our tables only refer to a standard Normal distribution; somehow, we must be able to transform the question about X to a question about Z.

Fig 9.17

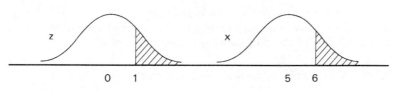

From figure 9.17, we can see that the area under the curve for X to the right of 6 is exactly the same as the area under the curve for Z to the right of 1. Thus, $P(X>6)=P(Z>1)$. All we have done is to take $P(X>6)$ and to subtract the mean value of X from both sides (to get $P(X-5>6-5)$) and then relabel $X-5$ as Z. In visual terms, all that has been done is that we have moved the curve for X five units to the left, so that it is now superimposed on the curve for Z. The result is that the line at $X=6$ is now superimposed on the line at $Z=1$.

Thus, $P(X>6)=0.1587$. Similarly, $P(X>3.1)=P(Z>-1.9)=0.9713$ and $P(4.6<X<6.3)=P(-0.4<Z<1.3)=0.5586$. Check these.

9.5.4 *Non-Standard Normal Distributions*

2.—Variance not equal to one.

Suppose that we have a Normal random variable Y with mean 0 and variance 4. What is $P(Y>2)$? In a similar way to the above case, we need to transform the question about Y to a question about Z.

In the previous case, the difference was one of location. Therefore, we shifted the curve. Here, the difference is one of spread. As a result, we "squash" the curve for Y until it takes the shape of the curve for Z. In fact, we divide both sides of the probability

Fig 9.18

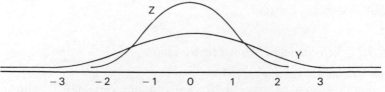

$$-3 \quad -2 \quad -1 \quad 0 \quad 1 \quad 2 \quad 3$$

question by the standard deviation of Y (*not* the variance). Thus, $P)Y > 2) = P[(Y/2) > (2/2)]$ and we then relabel $Y/2$ as Z. Thus, $P(Y > 2)$ is equal to $P(Z > 1) = 0.1587$.

Similarly, $P(Y < 1.6) = P(Z < 0.8) = 0.7881$ and $P(-1.64 < Y < -0.3) = P(-0.82 < Z < -0.15) = 0.2343$.   Check these.

(Note here that if the variance and, hence, the standard deviation, are less than one, our division by the standard deviation causes a "stretching" of the non-standard Normal distribution.)

9.5.5 Non-Standard Normal Distributions

3.—General Case

Here, we put together the two previous cases and consider a variable W with a non-zero mean and a non-unit variance. We shall use the general notation that $W \sim N(8, 9)$. This can be read as "W has a Normal distribution with mean 8 and variance 9."

To answer any probability questions about W, we merely "shift and squash", i.e. we subtract the mean of W and divide by the standard deviation (in that order).

Thus,

$$P(W > 9.8) = P(W - 8 > 9.8 - 8) = P\left(\frac{W - 8}{3}\right) > \left(\frac{9.8 - 8}{3}\right). \text{ We}$$

relabel the left hand side as Z. Previous theory enables us to show that $W - 8/3$ has a mean of zero and variance of 1.

Thus, $P(W > 9.8) = P(Z > 0.6) = 0.2743$. In general, any time we have a Normal distribution and we subtract its mean and divide by its standard deviation, we can relabel it as Z.

Similarly, $P(W < 10.58) = P(Z < 0.86) = 0.8051$ and $P(7.04 < W < 8.57) = P(-0.32 < Z < 0.19) = 0.2008$.

Example. The age distribution of buildings in a large industrial town can be adequately represented by a Normal distribution with a mean of 52 years and a variance of 256.

   (i) What proportion of buildings were built before 1900?
   (ii) What proportion were built between the two World Wars?

(iii) At what date had 22% of the present buildings already been built?

Let us take the present year as 1984, and let X be a random variable denoting age. Thus, $X \sim N(52, 256)$,

(i) $P(X > 84)$

$$= P\left(\frac{X - 52}{16} > \frac{84 - 52}{16}\right) = P(Z > 2) = 0.02275$$

(ii) $P(45 \leqslant X \leqslant 66)$

Here, we perform the same "shifting and squashing" to all three entries inside the brackets and then replace $X - \mu / \sigma$ by Z.

$$= P\left(\frac{45 - 52}{16} \leqslant \frac{X - 52}{16} \leqslant \frac{66 - 52}{16}\right)$$

$= P(-7/16 \leqslant Z \leqslant 7/8)$
$= 1 - P(Z \geqslant 7/16) - P(Z \geqslant 7/8)$
$= 1 - P(Z \geqslant 0.44) - p(Z \geqslant 0.88)$
$= 1 - 0.33 - 0.1899$
$= 0.4806$

(iii) Here, we wish to find k such that $P(\text{age} \geqslant k) = 0.22$. Now, we need to look at two other points first.

If Z equals $X - \mu / \sigma$ then X must equal $\sigma Z + \mu$

Also, if $k' = 0.77$, $P(Z \geqslant k') = 0.22$

$P(Z \geqslant k') = 0.22$
$P(\sigma Z \geqslant \sigma k') = 0.22$
$P(\sigma Z + \mu \geqslant \sigma k' + \mu) = 0.22$
$P(X \geqslant \sigma k' + \mu) = 0.22$

Now, all that is necessary is to evaluate $\sigma k' + \mu$ which equals $16 \times 0.77 + 52 = 64.32$. A rough answer then is 64 years ago or 1920.

Example. The prices (in thousands of pounds) of the properties on the list of one of Scotland's leading estate agents can be represented by a Normal distribution with a mean of 42 and a variance of 100. What proportion of these properties are under £29,000 in price? What proportion lie in the 40–50 thousand bracket? At what price do the top 4% of prices commence?

Let Y be a random variable denoting the price (in £000's)

$$Y \sim N(42, 100)$$

$$P(Y \leqslant 29) = P\left(\frac{Y-42}{10} \leqslant \frac{29-42}{10}\right)$$

$$
\begin{aligned}
&= P(Z \leqslant -1.3) \\
&= P(Z \geqslant 1.3) \\
&= 0.0968 \text{ or } 9.68\%
\end{aligned}
$$

$$P(40 \leqslant Y \leqslant 50) = P\left(\frac{40-42}{10} \leqslant \frac{Y-42}{10} \leqslant \frac{50-42}{10}\right)$$

$$
\begin{aligned}
&= P(-0.2 \leqslant Z \leqslant 0.8) \\
&= 1 - P(Z \geqslant 0.2) - P(Z \geqslant 0.8) \\
&= 1 - 0.4207 - 0.2119 \\
&= 0.3674 \text{ or } 36.74\%
\end{aligned}
$$

Find k such that $P(Y \geqslant k) = 0.04$

$P(Z \geqslant k') = 0.04$ if $k = 1.75$

$Y = 10Z + 42$
$P(10Z + 42 \geqslant 10k + 42) = 0.04$
$P(Y \geqslant 10 \times 1.75 + 42) = 0.04$
$P(Y \geqslant 59.5) = 0.04$

Therefore, a property price above £59,500 is in the top 4% on this list.

9.5.6　*Manipulation of Normal Distributions*

In this section, we shall deal with the general rules for simple combinations of Normal distributions.

Let $X \sim N(a, c^2)$ and $Y \sim N(b, d^2)$ and X and Y are independent of each other.

Then,

If $W = X + Y$, $W \sim N(a + b, c^2 + d^2)$
If $V = X - Y$, $V \sim N(a - b, c^2 + d^2)$ [Note: variance is not $c^2 - d^2$]
If $S = kX$, $S \sim N(ka, k^2c^2)$
If $T = X + h$, $T \sim N(a + h, c^2)$

Thus, we can produce some general rules such as

If $R = kX + h$, $R \sim N(ka + h, k^2c^2)$

and if we have several outcomes of the variable X denoted X_1, $X_2, \ldots, X_n$ then, if $Q = X_1 + X_2 + \ldots + X_n$,

$Q \sim N(na, nc^2)$ [Note: variance is not n^2c^2)

If M denotes the sample mean, i.e. Q/n, then $M \sim N(a, c^2/n)$

This is used frequently in later chapters.

These properties are valid provided each variable has a Normal distribution. Independence is required to include the variance. The expected values hold whether or not the variables are independent.

9.6 χ^2 Probability Distribution

We next discuss the χ^2 probability distribution (pronounced chi-squared to rhyme with sky). This is similar to the Normal distribution in that it is really a family of distributions indexed by two parameters.

The first parameter relates to the location and we shall always be dealing with what is termed the central χ^2. Non-central χ^2 can be dealt with in a similar way to dealing with Normal distributions with a non-zero mean.

The second parameter is called the "degrees of freedom" and relates to the number of observations we have. In most cases, if not all, that we discuss in this text, the number of degrees of freedom will be $n - 1$ where n is the number of observations. Now, for each number of degrees of freedom, the χ^2 distribution changes. In fact, a mathematical derivation of the χ^2 distribution with k degrees of freedom would be such that if $X_1, \ldots, X_k$ are independent random variables each with a $N(0, 1)$ distribution, then $X_1^2 + X_2^2 + \ldots + X_k^2$ has a χ^2 distribution with k degrees of freedom. A consequence of each degree of freedom resulting in a different distribution is that the tables generally list certain percentiles along each curve. Thus, a common table may, for a range of values of degrees of freedom, list values of W such that $P(\chi^2 \geqslant W) = 0.99$, 0.98, 0.95, 0.9, 0.8, 0.6, 0.4, 0.2, 0.1, 0.05, 0.02, 0.01.

9.6.1 Some Examples

We shall be mainly using the χ^2 distribution in later chapters especially for estimation and testing. However, it is important that we learn its use here.

If we refer to table 2 on page 264, we can answer the following questions.

What is the probability of a χ^2 variable resulting from 17 observations exceeding 29.633?

$$\text{17 observations—16 degrees of freedom}$$

$$\chi^2_{16,0.02} = 29.633 \text{ Hence, } P(\chi^2 \geq 29.63) = 0.02$$

What is the probability of a χ^2 variable resulting from 9 observations exceeding 11.03?

$$\text{9 observations—8 degrees of freedom}$$

$$\chi^2_{8,0.2} = 11.03. \text{ Hence, } P(\chi^2 \geq 11.03) = 0.2$$

If we have 11 observations contributing to a χ^2 variable, what is the 10th percentile?

$$\text{i.e. we want } k \text{ such that } P(\chi^2_{10} \leq k) = 0.1$$

$$P(\chi^2_{10} \geq k) = 0.9 = 4.865$$

These examples are unlikely to make you fluent in the use of the χ^2 distribution. However, they will ease some of the problems encountered when we begin to use this distribution in practice.

9.7 *F* Probability Distribution

The F distribution is another distribution which we shall be dealing with in later chapters. It is usually encountered when examining ratios of sample variances and is indexed not by one but by two parameters. These are both degrees of freedom; one for the numerator and one for the denominator.

When we have a ratio of sample variances, the variable will take a value

$$\frac{\Sigma(x_i - \bar{x})^2}{n_x - 1} \bigg/ \frac{\Sigma(y_1 - \bar{y})^2}{n_y - 1}$$

and the degrees of freedom on the numerator are $n_x - 1$ and degrees of freedom in the denominator are $n_y - 1$.

At the end of the text there appear tables of the F distribution. As in the X^2 distribution, a few percentiles are chosen and the corresponding values on a variety of F distributions are quoted in the tables.

For example, there is the (upper) 5% point of the $F(2, 3)$ distribution which equals 9.55 while the upper 1% point of the $F(8, 17)$ distribution is 3.79.

If we wish lower points, then we can use the fact that $F(\alpha; \nu_1, \nu_2) = 1/F(1 - \alpha; \nu_2, \nu_1)$ were ν is the Greek letter pronounced as gnu, the animal, and is generally used to denote degrees of freedom.

9.8 Conclusion to Continuous Random Variables

There are many other continuous random variables which are of interest and use in Statistics. These include the Uniform, the Exponential, the Gamma and the Beta. These will all appear in most texts in Mathematical Statistics or Applied Probability. However, there is really little use for them either now or in the short-term to medium-term future for members of your profession; only for the Statisticians.

As a result, we have dealt with only three distributions, and only one of these in any great depth. Do not minimise the importance of the other two. The brevity of their coverage in this chapter is due only to the fact that they are better understood when being used.

With this in mind, we can now progress to the next section of the text safe in the knowledge that we have, at the very least, a rudimentary understanding of our basic tool—probability.

Problems

1. If Z has a standard Normal distribution i.e. with mean 0 and standard deviation 1, what is the probability that
   - (i) Z is greater than 0.72
   - (ii) Z is less than 0.51
   - (iii) Z is greater than -0.6
   - (iv) Z is less than -1.33
   - (v) Z lies between 0.3 and 0.84
   - (vi) Z lies between -1.27 and -0.39
   - (vii) Z lies between -1.32 and 0.28
2. If Z has a standard Normal distribution, find a, b, c, and d, such that
   - (i) the probability that Z is greater than a is 0.14
   - (ii) the probability that Z is less than b is 0.37
   - (iii) the probability that Z is greater than c is 0.905
   - (iv) the probability that Z lies between d and $-d$ is 0.565
3. A random variable X has a Normal distribution with mean

25 and variance 25. What is the probability that X takes a value
   (i) above 35
   (ii) between 18 and 22.5
   (iii) between 22 and 30

4. If Y is Normally distributed with mean 6 and standard deviation 4, determine the values of a, b, and c, such that
   (i) the probability that Y is greater than a is 0.1
   (ii) the probability that Y is less than b is 0.61
   (iii) the probability that Y is greater than c is 0.97

5. The angle of elevation is measured from a fixed point on the ground to a fixed point on the side of a roof of a building. Over many years, this angle is measured by hundreds of students. It is known that the angles recorded by the students have a Normal distribution with a mean of 17.6° and a standard deviation of 0.8°. What proportion of students record angles in excess of 19.2°? What proportion record angles between 17° and 18.3°?

6. Let X be the random variable denoting the length of life of a certain brand of car battery. It is known that X is Normally distributed with mean 1,000 days and a standard deviation of 100 days. What proportion of these batteries last for over 1,140 days? What proportion last for between 760 days and 925 days? The manufacturer wishes to set a guarantee level so that only 10% of these batteries will fail before this guaranteed time has passed. What should the guarantee level be?

7. A certain building material is sold in nominal 25 Kg bags. However, the weight of material actually put into the bags has a Normal distribution with mean 25.2 Kg and a standard deviation of 0.125 Kg. What proportion of bags are, in fact, underweight?

8. The specified minimum strength of concrete cubes is 24 Nmm^{-2}. It is desired that not more than 4% of the cubes should fail to meet this mimimum. The strength of cubes is known to have a standard deviation of 2.8 Nmm^{-2}. What mean strength is required for the 4% condition to be satisfied? What mean strength is required if the condition is tightened to 2%?

9. Returning to the problem in question 5, the equipment used to measure this angle has been changed. If the mean is unchanged and the recorded angles still have a Normal distribution, what is the standard deviation of the recorded measurements if we know that 3% of recorded angles are in excess of 18.5°?

10. A plant producing binder mix for a highway surface has a qua-

lity control procedure based on the percentage of the aggregate passing a number 4 sieve. The measured variable is found to be Normally distributed with mean 40 and standard deviation 4.6. If the measured variable falls outside the range 34 to 46, the process is halted and adjustments made.

What is the probability that the process will be halted when it is running to specification (i.e. mean = 40, st. dev. = 4.6)?

What is the probability that the process will be halted if the mean process level has, in fact, dropped to 36?

11. A solicitor's scale fee for a particular service can be adequately modelled by a Normal distribution with mean £210 and standard deviation £62. Five clients, who can be considered to be independent, have this service carried out on their behalf. What is the probability that the total revenue to the solicitor from carrying out this service for these five clients is less than £1,000?

12. From past experience, it is known that the number of man-hours which it takes to construct a certain type of brick garage has a Normal distribution with mean 18.3 and a variance of 12.25 i.e. a standard deviation of 3.5 hours. If workmen are paid £8.40 per hour for building this garage, what is the mean and standard deviation of the amount paid to a workman for building this garage?

If 6 garages are to be built, what is the probability that it will take more than 100 man-hours? What is the probability that it takes between 102 man-hours and 115 man-hours?

13. The amount of water, W (in m^3), pumped from a ditch in a minute $N(80, 15)$. Let T be the time we have the pumped switched on. Let TW be the total amount of water pumped in time T. What value of T is necessary so that $P(TW \geqslant 1,000)$ is at least 0.975? (Hint: suppose we know T. Solve for $P(TW = 1,000)$ as far as possible to get an equation for T.)

Consider a different problem. Suppose that we have a large number of pumps which we shall switch on for a minute each. Assume that they operate independently of each other—a rather unrealistic assumption. Let N be the number of pumps we use and let WN be the amount of water pumped by N pumps in a minute. What value of N is needed so that $P(WN > 1,000)$ is at least 0.975? (Hint: suppose we know N. Solve for $P(WN = 1,000)$ as far as possible to get a quadratic equation. Then think carefully.) (Further hint: the answers for T and N should be close but not equal. Clearly, N can only take discrete values while T can take any value on a continuous scale.)

PART FIVE

Pure Problems

CHAPTER 10

Collection of Data

10.1 Introduction

In most of what follows in this text, we shall be dealing with information, usually numbers. As many practical statisticians would say, we shall be "getting our hands dirty". In this chapter, we shall discuss where these data come from, how we can collect them and some of the problems which arise.

10.2 Experimental and Observational Data

In surveying, a number of data sets are of observational data. We cannot go out and cause things to happen; we must merely observe them. For example, there may be thirty streets in a housing estate. If we wish to value a house on one of the streets, we may look at the sale prices of recently-sold houses in the estate. We might ideally wish to have information on four recent sales in each street but it might be that there has been no sale of houses on some of the streets. To get an accurate valuation, we cannot sell the house next door to the one under consideration to see how much it sells for, we must take the information that we have.

Observational studies (or observational data) are particularly common in the areas of medicine and law. In medicine, if we are trying out a treatment for cancer, we cannot give people cancer. We must wait until cancer patients come along. In addition, they must be willing to try the new treatment and we cannot plan on having, for example, equal numbers of men and women. Similarly, in law, if we are examining whether there is a relationship between a defendant's background and their chances of being acquitted, we cannot force twenty people from lower class backgrounds and twenty people from upper class backgrounds to commit the same crime. One must observe the crimes and cases which occur.

However, if we are collecting data by carrying out a statistical survey or a census, e.g. for market research, for election forecasting, etc., we can design procedures to be used and will know how accurately we are estimating parameters of interest such as a mean or a proportion. In the same way, if a surveyor is looking at a property, (s)he does not take any twenty pieces of information. Rather, the surveyor will examine specific aspects of the building and grounds and, from the information obtained, make a calculated judgement.

The rest of this chapter will be dealing with the procedures for designing a survey in the statistical sense of the word. Observational studies will not be mentioned as the analysis is difficult. Indeed, it is only recently that statisticians have begun to develop techniques of analysis.

10.3 The Benefits of Sampling

Let us suppose that we are trying to ascertain the mean disposable income for the working population of a large city. We could ask every single member of the working population about their disposable income. This would be called a 100% sample or, alternatively, a census. However, in most cases, we do not carry out a census. Instead, we take a sample. This is done for various reasons.

Cost: Although a sample does not give completely accurate information, we may have a sufficient degree of accuracy with a sample of 5% of the population or even less. Also, if we have a small sample, then we only need a few interviewers who can be well instructed and, more importantly, consistent.

Speed: The fewer people we sample, the quicker we have the analysis. Also, the fewer people we sample, the less time there will be between the first and the last person being questioned. This allows us to assume that the data were collected at a single point of time. (This is particularly important where opinions are of interest as opinions change with time.)

Equipment: In certain investigations, there is a need for a special piece of equipment to analyse the data or for a specialized person to elicit the information. Taking a small sample makes the investigation possible.

Destruction: There are certain situations where the act of sampling an item changes its nature. For example, testing sticks of dynamite by exploding them is very effective but if we test every one in this way, there are none left to be used. Similarly testing the lifetime of light bulbs by leaving them on until they fail is not very illuminating (sic). A third example would be in market research where we ask "Have you ever heard of Woofo dog biscuits?" They may not have before but they have now. We can call this destructive sampling as the nature of the sampled object is altered. In the last case, however, there may be a beneficial by-product for the makers of Woofo as their name becomes known.

10.4 Statistical and Non-statistical Sampling

Statistical sampling is a fairly objective approach in which a specified method of selection is specified before seeing any of the items to be sampled. Non-statistical sampling will be discussed later but it involves a number of subjective factors, some of which are advantageous and some are disadvantageous.

However, in either case, we may be interested in either attributes or opinions or both. By an attribute, we usually mean a measurement of some sort. For example, attributes of interest may include number of children, disposable income, age, number of rooms in a house, etc. By an opinion, we mean the same as the every-day use of the word. However, in most of what follows, we shall be dealing with attributes.

10.5 Statistical Sampling—some Definitions

Population: A population is a group of people, objects, etc., about which information is desired. It includes all of the people or objects whose attributes or opinions are relevant to a specific problem.

Sampling unit: Each individual member of the population is a sampling unit. Thus, a sampling unit can be a person, a house, a family, a warehouse, etc. It is always, however, the smallest sized member with which we deal.

Sampling frame: A sampling frame is a list of all the sampling units which together comprise the population.

Sampling scheme: A sampling scheme is a method or a set of rules for selecting sampling units.

10.5.1 Statistical Sampling—some more Definitions

Suppose that we are interested in estimating the mean disposable income of the working population of London. One approach might be to select buses as they pass through London and to board them and interview some of the passengers. Another approach might be to select a few people from the electoral register and to contact them and interview them. How do these approaches compare?

The first approach will tend to identify less affluent members of society. Thus, if we use the mean of the responses as an estimate of the mean disposable income for London, we will tend to have an under-estimate. This effect is termed bias.

Bias: A procedure is said to exhibit bias or to be biased if there is a systematic discrepancy between an estimate and the true value which it is supposed to represent. Bias can be established even without knowledge of the true value.

The second approach will probably not exhibit bias. However, by chance, it may provide a very low estimate or a very high estimate depending upon which electors are chosen. One measure of this variability is obviously the variance and, in most cases, we can evaluate approximately the variance of all the possible estimates of the mean disposable income of London.

In general, we wish a sampling procedure which is unbiased and has a low variance. However, to be both, the procedure will be expensive in terms of time and money. As a result, some compromise is usually reached.

We must now consider random samples. A random sample is such that every possible choice is equally likely. For example, if we were to select six people from a group of 30, there are 593,775 possible selections. Random sampling dictates that each of these 593,775 possible selections is as likely as any other. Thus, the term random is not applied to the group of six people actually chosen. Rather, it is applied to the process by which the group was selected.

Having looked at bias, variance, and randomness as applied to sampling procedures, we can now proceed to consider some sampling schemes. To make the problem more concrete, we can consider the problem of selecting a sample of size twenty from the population of registered estate agents in the UK.

10.6 Simple Random Sampling

A simple random sample is a random sample which is unrestricted. If we have the sampling frame, i.e. a list of all registered estate agents, then a simple way is to number the estate agents. We can then select random numbers to identify the twenty estate agents chosen. (For random numbers, see 10.10.)

This procedure may or may not be unbiased depending upon what we are trying to estimate. However, it usually has a high variance relative to other methods of selecting a sample of size twenty but is simple to operate.

10.6.1 Stratified Random Sampling

A stratified random sample is a random sample which has restrictions placed upon it. We must first divide the population into layers or strata. We can divide the estate agents according to several variables. We can divide them by annual turnover or by number of outlets or by length of operation or by number of partners or by several other variables. Which variable we use may depend on the availability of information and on relevance to what we are trying to estimate. However, as an example, we shall consider the estate agents to be divided by turnover.

One grouping of estate agents may be to have the 15% of estate agents with the largest turnover to be grouped in stratum A. Stratum B will contain the next 25% of estate agents, while stratum C will contain the 60% of estate agents with the lowest turnover. Now, instead of selecting an unrestricted sample of twenty estate agents, we can choose how many to select from each stratum. One approach, although not necessarily the best, is to select three from stratum A, five from stratum B, and twelve from stratum C. This is proportional allocation, i.e. allocating sample size to each stratum proportional to the stratum's size as part of the whole population. Proportional allocation is often used but it is not always the best method of allocating sample sizes to each stratum.

What is happening is that we are selecting a simple random sample of size three from stratum A, a simple random sample of size five from stratum B, and a simple random sample of size twelve from stratum C. The benefit of this method is that we are assured of the inclusion in the sample chosen of estate agents across the spectrum of turnover. Simple random sampling may choose twenty estate agents who all fall in stratum C, or even all in stratum A. If we were interested in estimating mean turnover, then we would be in danger of producing an estimate greatly in error. The cost of this method is that we need full information on a relevant variable by which to stratify the population. Whether the benefit is greater than the cost will depend upon the exact purpose of the study and the accuracy required. However, some sort of stratification is usually carried out as it often cuts the costs of administration and the management of a survey.

10.6.2 Systematic Sampling

Here we must take the sampling frame and put the estate agents in some sort of order. If we suppose that there are 4,000 estate agents in the UK, then we wish to select one from each 200 on the list. To do this, we randomly select one from the first 200 and then every 200th one after that. This gives a systematic sample of size twenty. Note, however, that this is not a random sample. Each estate agent has the same chance of being selected but it is *not* the case that each possible selection of twenty estate agents is equally likely.

How good a method this is often depend upon the variable we use to order the estate agents. If we order the estate agents by turnover then, as in a stratified random sample, we shall be choosing estate agents across the spectrum of turnover.

If the estate agents are ordered alphabetically, as they might be in a national register or a national telephone directory, then nothing would be gained. If they are ordered by the number of outlets or agencies, then this may be helpful as this will be correlated with turnover.

This method of sampling has a number of advantages depending upon the situation. It is very simple in some general outdoor situations where we can sample every tenth person passing a street corner or reaching a toll at a bridge. If we do this, we do not require knowledge of the sampling frame although the analysis may be hindered as a result.

There are disadvantages, however. If we choose every kth item and k is chosen badly, we run into problems. For example, if we choose every sixth day of a shop's daily returns, then we are able to estimate the turnover for Thursdays, say, but have no information on the other days on which the shop is open. This is only an example of the way in which we may pick up a cycle and, as a result, lose valuable information.

Further, if k is not a factor of the total number of the population, i.e. does not divide exactly into it, then other problems are encountered. Suppose we wish to select every fourth unit from 18. If we choose 1, 5, 9, 13 and 17 or 2, 6, 10, 14 and 18, we have a sample of size 5. Choosing 3, 7, 11 and 15 or 4, 8, 12 and 16 gives a sample of size 4. Thus, our sample size may depend upon exactly which sample is chosen.

Other practical problems occur but we shall not bother with them in our discussion of the ideas of different sampling methods.

10.6.3   Cluster Sampling

With stratified random sampling, we divide the population into strata, select a few from each stratum and this will mean that we have, hopefully, a fair representation of the population in our sample. Cluster sampling, in some sense, takes the opposite approach.

Instead of dealing with strata which are internally similar, we deal with clusters which, hopefully, are internally dissimilar and, in fact, mirror the population. Thus, in our example, we could take geographical areas as clusters. Instead of choosing twenty estate agents in the UK we can choose one cluster, i.e. one area, for example, West Midlands, and choose twenty estate agents from this area. What we are assuming, then, is that as far as estate agents are concerned, the West Midlands is a microchosm of the UK; that the West Midlands exhibits the same or similar peaks and troughs as the UK as a whole.

The advantages of cluster sampling are that it may reduce time and travelling costs as well as requiring a sampling frame for only one or a few regions. The disadvantage is that we must make the assumption that the clusters do, in fact, mirror the population. When sampling, we only collect information from one or a few clusters and we cannot verify our assumption.

10.6.4 Double Sampling (or two-phase sampling)

The principle of double sampling is that we actually take two samples. It is particularly useful when we know very little about the population being sampled. For example, we may feel that a stratified sample ought to be employed but do not have any knowledge about the strata. Another example is when we wish to have the variance of an estimator to be within certain bounds but do not know how large a sample to take. As a result, this method is particularly useful when there is a limited amount of money to be spent.

The first sample taken is generally a relatively large unrestricted simple random sample which provides the desired information. Using this information, we can carry out a smaller more restricted second sample. For any analysis, the data from the two samples must be pooled.

10.6.5 Multi-Stage Sampling

Unlike double sampling, multi-stage sampling only involves taking one sample. However, it involves two or more stages in selecting the units. For example, in selecting estate agents, we could proceed as follows:

(1) Divide UK into regions—clustering
(2) Divide regions into rural/urban
    Select two of each type—stratification
(3) Obtain/compile a local register for these regions. Choose

every *k*th estate agent to select twenty agents—systematic sample

Multi-stage sampling can be useful as it reduces travelling costs, interviewing costs, administration etc. However, unless we have full information about the whole population and its strata and clusters, we are unlikely to be able to evaluate how accurate our results are.

10.6.6 *Sequential Sampling*

In all of the sampling schemes we have discussed, at some point, a decision has to be made to sample *k* sampling units. It is also the case that we collect all the information before carrying out any analysis. This general approach is fixed sampling.

An alternative approach is sequential sampling. Here, the number of units sampled is not specified beforehand. Instead, we establish some criterion or decision rule and continue sampling until this criterion is satisfied.

For example, we may state that we shall continue to sample estate agents until we have at least four with turnover in excess of £1m and at least six with turnover less than £200,000. Note that this is not the same as sampling four over £1m and six less than £200,000. If it takes sixteen, say, less than £200,000 before we have four in excess of £1m, this provides information on the relative frequencies of these two sizes of turnover, i.e. that there are about four times as many below £200,000 as above £1m.

In general, for a given reliability or variance, sequential sampling will tend to lead to smaller sample sizes. However, the analysis is far more difficult. In addition, it may be less expensive to select and analyse a large sample at one time than to draw units one (or a few) at a time and analyse them before continuing.

10.6.7 *Area Sampling*

Most, if not all, of the above sampling schemes require knowledge, at some point, of a sampling frame. However, in certain situations the sampling frame is either impossible or too expensive to obtain or is too inaccurate. In these cases, a common solution is to resort to area sampling. The fundamental idea behind area sampling is that the sampling units of almost any population, e.g. shops, homes, solicitors' practices, etc., can be identified in terms of geographical location and these geographical locations can be listed. Thus, the units are sampled indirectly. An example of area sampling comes from a survey of retail distribution carried out in the USA.

"The United States is composed of 3,136 counties and parts of counties; these have been grouped into 1,893 primary sampling units ... The 1,893 sampling units are grouped into 125 strata ... One primary sampling unit is selected within each stratum on the basis of probability proportionate to retail population ... The next step is to select a sample of small nonoverlapping land areas within the primary sampling units ... All retail outlets are enumerated within these clusters ... The sample of stores which are contacted in an audit is drawn from outlets which were found in the sample census to be carrying the types of products."

As can be seen, both strata and cluster are mentioned here. To obtain an area sample, some grouping of sampling units is necessary. These groups are not thought to be internally similar or dissimilar but are merely to get one stage of the sampling procedure down to a practical size.

The statistical formulae relating to area sampling are cumbersome. In general, however, the error due to area sampling is greater than in less restrictive samples which take units over a wider area but require a more extensive sampling frame.

10.7 Panel Surveys

Panel surveys do not constitute a different method of selection. Rather, it is what we do with those units selected for. In a panel survey, we sample a group of people, companies, households, etc. and obtain information from them, not once, but at several points in time.

Panels are most commonly used:

(1) as consumer purchase panels, where people periodically record and submit the purchase they make;
(2) as trade panels, where dealers regularly submit information on levels of sales, prices and inventory; and
(3) as opinion panels, where people are, from time to time, asked their opinions of, for example, government policy.

Panels are clearly of use as they greatly reduce the cost of repeatedly selecting sampling units. However, certain problems do arise. There is a significantly high rate of refusal to participate in a panel survey and there is a high drop-out rate once the survey has begun. Both of these are serious problems as there is strong evidence that the characteristics of those who refuse and those who drop-out are different from those who agree and continue to participate.

Thus, a degree of bias may be introduced and an adjustment will need to be made to the results.

Another problem is that continual participation may have an effect on behaviour or opinion. Panel members may become atypical in their behaviour or habits as a result of being continually under scrutiny.

The advantages of panel surveys are that they reduce costs and reduce the time involvement necessary to obtain information repeatedly. Also, they are good for "before-and-after" studies, i.e. for assessing the effects of, for example, a change in the tax structure or a change in the rôle of building societies.

10.8   Non-Statistical Sampling

Although non-statistical sampling is subjective, there are a number of situations where it is of great use. In this section, we shall discuss four types of non-statistical sampling. However, it ought to be noted here that these four types are not necessarily distinct; rather, they are four different aspects.

10.8.1   Judgement Sampling

If we are selecting a sample of twenty estate agents, it may be that it is felt that two or three large estate agencies must be included to give a useful, accurate, and credible result. In other words, there is a subjective assessment of units to be selected.

A judgement sample is used in certain situations when the only basis for inclusion or exclusion in a sample is the judgement and experience of an expert. Such situations are troublesome, however, because of the inability to assess the accuracy of the expert and because of the difficulty in replacing experts.

10.8.2   Snowball Sampling

This is a form of sampling which is appropriate when the objective is to reach a small specialized population. For example, we may be interested in questioning surveyors who are actively involved in using micro-computers in their work. This group may be employed in diverse organizations and live in various parts of the country. The idea of snowball sampling is to contact a few of these people and, as well as asking them for the information you seek, ask them to give you the names of two or three more people for you to contact. The result is that each person sampled identifies others and means that there is no need to construct a frame. One problem is that those who are more visible, either socially or professionally, are more likely to be sampled.

10.8.3 Convenience Sampling

To obtain information and opinions cheaply and quickly we should contact the most convenient units. The most convenient people to sample are friends, relations, fellow employees, and volunteers. In marketing research, this type of sampling is often carried out at a very preliminary stage, particularly to test questionnaires before being distributed on a large scale. However, to use this type of sampling as a sole means of selecting units is, in general, indefensible.

10.8.4 Quota Sampling

This is the most common type of non-statistical sampling and, in concept, is a hybrid of judgement sampling and stratified sampling. The population under consideration is split into several groups and we then decide how many of each group is not carried out as in statistical sampling. In most cases interviewers will be sent out to collect the information. Suppose, for example, that the population is divided by age. The categories might be 18–29, 30–44, 45–59, and over 60. An interviewer might be instructed to ask 100 people between the ages of 18 and 29. The interviewer chooses which people to approach and, obviously, the first question is concerning the person's age. If their age is within the 18–29 range, subsequent questions will be put. This is a quick and relatively easy way to select samples.

However, mistakes will occur in selection. Obviously, to cut down on mistakes, the interviewer will tend to stop only those people who look as if they are well inside the range. As a result, most of the people selected will have ages between 21 and 25. Thus, we will not have ages spread evenly over the whole range.

This method, as was stated above, is quite common particularly in sampling people on the street. It does not need a frame, is fairly low in cost and takes relatively little time. However, the problems occur mainly in the lack of objectivity, about which much has already been said.

10.9 Sources of Error

Sampling is a step in the collection of useful information. However, the ideas of sampling may provide some light on the types of errors and inaccuracies which occur in the information collected. The sources of error detailed below do not apply to all types of survey, but are, each of them, quite common in various contexts.

10.9.1 Sources of Error—Inaccuracy

There are many sources of inaccuracy when we carry out a survey. Questions, either verbal or written, may be ambiguous. Thus, people's responses may not be to identical questions. The answers could be recorded wrongly. This could happen if an interviewer makes a mistake in writing down a reply or if some instrument is used to measure some variable and the instrument is imprecise. A third possibility is using published data which are out of date. Further, if an interviewer is involved, there is a possibility that the interviewer will misinterpret the response.

10.9.2 Sources of Error—Lack of Truth

People invariably fail to tell the truth. This is particularly true of age and income of people and of companies' turnover or property owned. In cases such as these, the best approach is to seek another source of information to verify the survey's data. For example, we may check the register of births or, perhaps, get the Inland Revenue to assist or examine registered accounts or the companies' register.

What is equally important as verifying the data is examining the reasons for failing to tell the truth. In some cases, it may be that a full completion of the survey requires a great time commitment. As a result, people may tend to rush through it either answering with their first impressions or with any answer at all, or failing to get accurate information giving approximate answers only. In many cases, answering questions truthfully involves either a loss of prestige or an invasion of privacy. This loss and/or invasion could be either of a personal form or in business. Another reason for receiving less than the truth is that a person may alter his/her opinions to more closely match those of the interviewer. One way to avoid this is to train interviewers not to project their own personality and opinion while attempting to elicit opinions from interviewees.

10.9.3 Sources of Error—Bias

We have previously discussed bias but, in most sampling situations, we need to discover its existence, or likely existence, and to try to correct for it by estimating how much bias there is. In some situations, methods do exist to counteract any bias but we shall not discuss them here. What is important is that we try to discover whether there is any bias.

10.9.4 Sources of Error—Non-responses

It is only recently that the subject of non-responses has received much attention in the statistical literature but this subject is of great importance. Questions are often not answered and it is vital that we consider the reasons for the questions not being answered.

One possible reason is that of non-coverage. For example, if a selection of possible answers is provided, it may be that the answer you wish to give is not one of the possible answers. This is clearly caused by poor survey design.

Another possible reason is that, despite repeated efforts, a selected person has not been contacted.

Also possible is that someone is unable to answer a particular question. This is usually caused by them not having the required knowledge.

A further possibility is that someone refuses to answer a question. This will happen for the same reasons as people will lie.

People may, of course, omit to answer certain questions. This is usually caused by such simple things as rushing or simple forgetfulness.

The last common sources of non-response is the "hard core" of people who refuse to answer questions. Unfortunately, there is little we can do about these people.

The effects of non-response are varied but two main ones ought to be mentioned. Firstly, if we wish a sample of size 200, say, then, once we suffer the non-responses, we may only have 170 or 180 responses and so the effective sample size has been reduced. Thus, if we require 200 responses for the sample of a particular degree of accuracy, we generally estimate the non-response rate and inflate the sample size accordingly. Thus, we might sample 240 people so as to obtain 200 responses. However, this involves us in extra cost.

The other main effect is bias. This occurs because, in many situations, it is quite reasonable to assume that those who fail to respond, for whatever reason, are different in character than those who respond. In other words, were we able to elicit a response from the non-respondents, these forced responses would be different, on average, or would give us different results from the responses of those who actually responded.

There are three basic ways to treat these non-responses. The first one is to ask them the questions again. This will, hopefully, remove those who omitted to answer. In addition, we can try to record the answers of those who fall in the non-coverage category as well as trying to discover the reasons why some others did not answer.

Secondly, as in dealing with a failure to tell the truth, we can try to fill in the missing information from other sources. This may involve us in much extra time and expense and so we must consider whether the benefit outweighs the cost.

The third way is to use other information to build up a mathematical model to try to predict what the non-respondents would have responded. This approach has generated a lot of activity recently in the statistical world but is not yet of proven value to the practising non-specialist.

10.9.5　Sources of Error—A Conclusion

It is impossible to collect data without error. All we can do is to try to identify sources of error and to minimize the chances of these sources generating errors. The important thing is that we are aware of the sources of error, the types of error and their magnitude and can, therefore, adjust for them. More than this is, at present, too much to ask.

10.10　Random Numbers

We often need a numerical method of selecting a random sample. If the total population is small, we can write the names of each member of the population (or an identifying number) on a piece of paper, roll them up, place them in a bag, shake up the bag and select the requisite number of pieces of paper. The problems with this are that it is cumbersome, takes up a lot of time and paper and relies on the rolled up pieces of paper being of roughly the same size and weight. Also, if the population is large, this method is impractical.

Other similar methods are possible such as using playing cards or dice but the most useful method is the use of random numbers or random digits. As was stated earlier, the term random does not really apply to the number, rather it applies to the process by which the numbers were generated. However, we do not need to concern ourselves with producing these random numbers as most statistics books contain at least one page of these numbers. What is important is that these digits have been tested so that they satisfy the criteria we would wish, i.e. the ten digits 0, 1, 2 ..., 8, 9 appear with roughly equal frequency, without pattern, and the sequence of digits appears to be independent.

Random numbers are often printed in columns of two or of five but that is merely a style of printing. They are to be considered as separate digits. To discuss the use of random numbers, let us consider a specific example. Let us suppose that we have a list of 73 building contractors and we wish to sample 10 of them. The first thing to do is to get the page of random numbers and select a starting point at random. One way to do this is to close your eyes and pick a point. However, an improvement on this is to pick a point and then select the next four digits reading across horizontally. (If you reach the edge of the page, move on to the next row.) Take these four digits and consider the first two as a two-digit number and the second two as another two-digit number. Move the starting point down by the same number of rows as the first two-digit number and then move it across by the same number of columns as the second two-digit number. (If you reach the bottom of the page, move to the top of the page.) Now we have a starting point which is fairly close to being random.

Suppose that a portion of the table of random numbers is as follows:

03925	61843	95582	80433
97256	78380	00127	61488

Our starting point will be the 3 at the end of the 61843 group. We have the contractors numbered from 01 to 73. The first two-digit number is 39 and so we select contractor number 39. Then, we select contractor number 55. The next two-digit number is 82. This is not in the range 01 to 73 and so we discard it. We also discard 80. Then. we select contractor 43. The next two-digit number is 39. However, in most cases, we do not wish to sample a sampling unit twice. Thus, we discard this number. Then, we sample contractors numbered 72, 56, 78 and 38. The next two-digit number is 00. This also is not in the range 01 to 73 and so is discarded. We then select contractors numbered 01, 27 and 61 and stop as we have now selected 10 contractors, i.e. those numbers 39, 55, 43, 72, 56, 38, 01, 27, 61, and 48.

This process is not only simple but is quite adaptable. If we have 629 contractors in one sampling frame, then we take the digits three at a time and reject 630 to 999 and 000. In addition, one or two little tricks do exist to speed matters up but this general procedure is far faster than using pieces of paper or dice.

Random numbers appear in Table 4 on page 270.

Problems

1. Consider the problem of finding the views of your profession concerning a proposed change in policy or in administrative structure or in a code of practice. There are many ways to assess the opinion of your fellow professionals. Think of some of these and of the operational problems with each of them. What are you trying to measure and of whom? Should all people have the same weight in your considerations? What biases can be expected to be introduced unless you plan to counteract them?

2. Your organisation wishes to construct its own House Price Index. (For index numbers, see Chapter 17.) What sort of data would you require, what data are available, and how well do these match? What problems are likely to occur?

3. How would you assess customer satisfaction in your profession? Who exactly are your customers? In what ways might you collect information? What problems are likely to occur?

CHAPTER 11

Estimation and Sampling—Properties

In this chapter we shall briefly examine three of the properties and concepts associated with sampling and estimation.

11.1 Robustness and Efficiency

At various points in this text, either implicitly or explicitly, we shall make assumptions in carrying out a procedure. When doing this, we shall often be making a decision between employing an efficient procedure or a robust procedure. It is important that we understand these two terms.

An efficient procedure is one which makes full and good use of the data available and of our knowledge or assumptions about the underlying probability distribution from which we are sampling. A robust procedure is one which does not rely too heavily on the underlying assumptions which are generally about the probability distribution from which we are sampling.

For example, in Chapter 5 we discussed the mean and the median as summaries of location. The mean is an accurate summary of location but is only accurate if the data are approximately symmetric about the centre. However, if there is this rough symmetry, then the mean uses all the information and is a summary of location. If the data are significantly asymmetric, i.e. if the data are significantly skewed, then the mean will be a very inaccurate summary of location.

The median is also an accurate summary of location and does not rely for its accuracy on such a strong assumption. The median does not use all the information but will always be a fairly accurate summary.

In summary, the mean is an efficient summary of location which, when the symmetry assumptions holds, is very accurate but may be very inaccurate. The median, however, is a robust summary of location which will always be reasonably accurate.

We might say that using a robust procedure is a conservative approach while using an efficient procedure is a bolder approach. Classical statistics has gone for efficiency and in most of this text we shall follow this line. However, it is important that we understand the distinction.

11.2 The Central Limit Theorem

The Central Limit Theorem is one of the fundamental theorems of statistics. It states that:

For (almost) any population with finite mean μ and finite variance σ^2, if we take a random sample of size n, where n is sufficiently large, and compute the sample mean $\bar{x}$ and repeat this process several times, then these sample means will follow an appropriate Normal distribution, the approximation being closer the larger n is, this Normal distribution having mean μ and variance σ^2/n.

This theorem is important as we now have information about the distribution of the sample mean irrespective of the distribution of the population from which we are sampling.

As an example, let us look at the problem of trying to estimate the mean salary of a large English town. Suppose that we are interested in the working adult population of the town and that we took a random sample of size 100 which had a mean of £9,384 and a standard deviation of £630. One thing that we know for sure is that the mean for the whole town is not exactly £9,384—but it may be close. In fact, by the Central Limit Theorem, we know that this sample mean has an approximate Normal distribution with mean μ and variance σ^2/n.

Thus, we may be able to assume that if the sample variance is reasonably accurate, the standard deviation of the sample mean is close to $\sqrt{(630)^2/100}$ which equals 63. Thus, we have some idea of how far the mean is likely to be from our sample mean of £9,384. We shall be utilizing this theorem and this approach considerably in subsequent chapters, especially Chapter 12.

11.3 Finite Population Correction

When involved in sampling exercises, we sometimes are dealing with populations which are so large that we can consider the population size as infinite. However, this is not always the case as we are often dealing with small samples which must be considered finite. When we have a finite population of size N, say, we need to make a finite population correction when evaluating the variance of the sample mean. This is because, if we sampled all N units, the variance would be zero as we have all the information and can, therefore, calculate the mean exactly.

If the population size N and the sample size is n, the variance of the sample mean is

$$V(\bar{X}) = \frac{\sigma^2}{n}\left(1 - \frac{n}{N}\right)$$

where the $(1 - n/N)$ expression is known as the finite population correction or f.p.c.

Your can see that if we take a census of this population, i.e. if $N = n$, then the variance is zero.

In most of the expressions in the chapter(s) below where σ^2/n is written as the variance of the sample mean, the f.p.c. ought to be introduced if the population has to be considered finite. Such a consideration is probably necessary if N is less than 1,000 or if we are sampling 5% or more of the population.

Suppose, for example, that we are trying to estimate the mean return on a three-year investment of £1,000. We are only considering building society fixed-term bonds and believe that the market for these is such that the variance should be about the same as it was six years ago when a thorough study was carried out by the BSA. (This stability of variance, but not of the mean, can be justified by an understanding of the market forces.)

Suppose that the previous variance for all such returns was 154. Thus, the standard deviation was £12.41. Now, in the present study, we are considering 74 bonds and have selected a sample of 14 of them. If we had a very large population, we would say that the variance of the sample mean is $154/14 = 11$ i.e. a standard deviation of £3.32.

However, we have quite a small population and so the f.p.c. must be called into action. Thus, the variance of our sample mean is

$$\frac{154}{14}\left(1 - \frac{14}{74}\right) = 8.919$$

i.e. we have a standard deviation of £2.99. This can be related to the mean return in our present study to give some idea of how accurately we are able to estimate the mean return for all 74 bonds.

Problems

1. A large life assurance company knows that, in a particular sector of its market, the time until maturity of their policies has a mean of 8.5 years and a variance of 8.3 (years²). If we take

a random sample of 80 policies from this sector, what is the approximate distribution of the sample mean time until maturity?

What is the probability that the sample mean is more than 9 years?

How large a sample is needed so that this probability is less than 0.03?

How large a sample is needed so that the standard deviation of the sample mean is less than one month (i.e. $\frac{1}{12}$ of a year)?

2. The rateable values in an area are known to have mean £389 and standard deviation £73. If we take a sample of 140 properties, what is the probability that the sample mean rateable value exceeds £400? What size of sample do we need to get a standard deviation of the sample mean of less than £5?

3. In a north of England city, the distance people travel to work is known to have a mean of 3.3 miles and a variance of 3 (miles^2). If we sample 120 people, what is the probability that the sample mean distance is less than 3 miles?

What is the probability that the sample mean distance is between 3.2 and 3.4 miles?

Suppose that we do not know the mean distance people travel to work in the city. Call it μ. Now, evaluate the probablity that our sample mean is within 0.1 of a mile of μ, i.e. is between $\mu - 0.1$ and $\mu + 0.1$. (The answer should not involve μ. This will become important later on when we try to estimate μ as we can now make probabilistic statements about our estimates without actually knowing the value we are trying to estimate.)

How large a sample do we need to make this probability exceed 0.9?

4. Suppose, in question 1 above, we are actually dealing with a very small sector in that there are only 710 policies. What now is the standard deviation of the sample mean?

5. Sensitivity analysis has recently become important in property valuation. A computer consultant is testing out the available software and is interested in the prices of these packages. If there are 34 packages currently available, and he samples seven, and the variance of the prices of all 34 is 1,724 (£^{2}), what is the standard deviation of the mean price of his sample of seven?

CHAPTER 12

Estimation

12.1 Introduction

In this chapter, we shall be examining one of the basic problems in Statistics. This problem is one of estimating some characteristic or parameter of a population.

As has been discussed in the previous chapter, we generally take a sample and, because of this, any estimate we make will depend on the accuracy and appropriateness of the sampling procedure used. In addition, it must be realized that, if a second sample were taken, we would almost certainly have a different set of results and so we could produce a different estimate.

For example, suppose that we are trying to estimate the mean wage for unskilled workers in the UK. One researcher could go out and contact 40 randomly chosen employers, say, and obtain the relevant information. Provided his choice of employers is unbiased and the information is accurate, one would expect the mean from this sample to be close to the true mean of the population of all unskilled workers in the UK. However, another researcher could carry out the same procedure and pick another random sample of 40 employers. There may be some overlap between the two sets of employers. However, the second researcher will have different figures and, as a result, will obtain a different sample mean to use as an estimate. Which one is correct?

In truth, neither are strictly correct. Suppose the first researcher has an estimate of £104.43. It is unlikely that the mean of the population is exactly £104.43. What is far more informative is to reveal how likely the mean is to be below £104, £103, £100, or £95, on the basis of this sample. What is the second researcher has an estimate of £107.61? We may suggest that the population mean is quite likely to lie between these two but is £108.09 a likely population mean or £110? How likely is £104 now? or £95? For this reason, we do not produce a single estimate—which could be called a point estimate. Rather, we produce a range of likely values for the mean of the population. This is called an interval estimate.

In this chapter, we shall construct interval estimates for several quantities in different situations.

12.2 Estimating a Mean

12.2.1 *Estimating a Mean Salary*

To continue with the previous example, we have one researcher who collected data on 740 unskilled workers. The mean of their salaries was £108.32 and the standard deviation of their salaries was £9.37. From this we wish to estimate the mean salary of all unskilled workers.

To do this, we have to use the Central Limit Theorem. If we denote the mean of the sample by $\bar{X}$ then, if the salaries of all unskilled workers have a mean of μ and a variance of σ^2, we know that, by the CLT

$$\bar{X} \div N\left(\mu, \frac{\sigma^2}{n}\right)$$

We do not know μ or σ^2 but hopefully £108.32 will be close to μ.

Suppose we consider a range of likely values for μ and say that we wish to be fairly sure that this range of values actually contains the unknown population mean μ. For the moment, let us quantify "fairly sure" by saying that we want to be 95% sure that μ lies in this range of values.

If we have a standard Normal distribution, then we know that

$$P(-1.96 < Z < 1.96) = 0.95$$

Recalling that $\dfrac{\bar{X} - \mu}{\sigma/\sqrt{n}}$ can be expressed as Z, and vice versa,

$$P\left(-1.96 < \frac{\bar{X} - \mu}{\sigma/\sqrt{n}} < 1.96\right) = 0.95$$

Multiplying all three terms by $\sigma/\sqrt{n}$ gives

$$P\left(-1.96\frac{\sigma}{\sqrt{n}} < \bar{X} - \mu < 1.96\frac{\sigma}{\sqrt{n}}\right) = 0.95$$

Subtracting $\bar{X}$ from all three terms gives

$$P\left(-1.96\frac{\sigma}{\sqrt{n}} - \bar{X} < -\mu < 1.96\frac{\sigma}{\sqrt{n}} - \bar{X}\right) = 0.95$$

Multiplying by -1, which requires reversing the inequalities, gives

$$P\left(\bar{X} - 1.96\frac{\sigma}{\sqrt{n}} < \mu < \bar{X} + 1.96\frac{\sigma}{\sqrt{n}}\right) = 0.95$$

*

i.e. if $a = \bar{X} - 1.96\dfrac{\sigma}{\sqrt{n}}$

$b = \bar{X} + 1.96\dfrac{\sigma}{\sqrt{n}}$

then the range of values from a to b is a likely range of values for μ and we call the interval from a to b written (a,b) a 95% confidence interval.

In this case, we do not know σ. However, we do have an estimate of σ in the standard deviation from the sample. If we assume that we have a large sample and that our estimate of σ will be accurate, we can construct a 95% confidence interval by calculating

$$a = 108.32 - 1.96 \times \frac{9.37}{\sqrt{740}}$$

$$= 107.64488$$

and

$$b = 108.32 + 1.96 \times \frac{9.37}{\sqrt{740}}$$

$$= 108.995118$$

Thus, (£107.64½, £108.99½) is a 95% confidence interval for the mean salary of all unskilled workers.

There is one important point to be noted here. Equation * above is not strictly correct. The numbers are correct but it is not, strictly speaking, a probability. In the case above, we have an interval from £107.64½ to £108.99½. It is not strictly correct to say that the probability that the mean in this interval is 0.95. Either it is in the interval (with probability one) or is outside it (with probability one). To be more accurate, we would say that if the sampling process was repeated by several researchers, about 95% of the intervals which were constructed could contain the mean wage of all UK unskilled workers.

Example. A life assurance company samples 140 of their life policies and records how long these policies have still to run until they mature. The sampled policies had a mean of 7.1 years and a standard deviation of 1.86 years. Construct a 90% confidence interval for the mean time until maturity of all their policies.

This problem is basically the same as the previous one. However, we now are going to construct a 90% confidence interval. From the Normal distribution, we use 1.645 instead of 1.96. Further, we again assume that the standard deviation of the sample is a fair estimate of the standard deviation of the population.

Thus,
$$a = 7.1 - 1.645 \times \frac{1.86}{\sqrt{140}}$$

$$= 6.8414$$

and
$$b = 7.1 + 1.645 \times \frac{1.86}{\sqrt{140}}$$

$$= 7.3586$$

and the 90% confidence interval for the mean time to run is (6.8414 years, 7.3586 years).

12.2.2   What Level of Confidence do we wish?

In the last example, if we wanted to be 100% confident, we either sample every policy or have an interval which goes from 0 to infinity. The more confident we wish to be, the wider an interval we need. The wider the interval we have, the less use it is. However, all is not despair. Firstly, with a reasonable sample size, we can produce an interval of moderate width which has, attached to it, a fair degree of confidence. Secondly, an interval is still far more useful and informative than a point estimate.

12.2.3   How Large a Sample do we need?

In general, if the sample size is at least thirty, the Central Limit Theorem provides a very good approximation for the distribution of the sample mean.

If we have less than thirty observations, then the CLT does not provide a particularly good approximation. In fact, if we have a small sample, we do not use the Normal distribution but instead use the Student's t or t distribution. Details of this distribution can be found in many statistics textbooks.

12.3 Estimating a Proportion

There are 10,000 homes in a neighbourhood. One hundred are sampled and 40 of those sampled have double glazing. Double glazing companies wish to estimate the proportion of the homes in the neighbourhood who already have double glazing.

Once again, we could produce a point estimate. This would be 0.4. However, this is of little use. It will be much more useful to quote an interval—a confidence interval.

Denote the time proportion which we are trying to estimate by π and our estimate by p.

The Central Limit Theorem applies and, provided π is not too far from 0.5,

$$p \doteqdot N\left(\pi, \frac{\pi(1-\pi)}{n}\right)$$

where n is the sample size. (π is usually judged to be "too far" if it is less than 0.1 or greater than 0.9).

We can use this distribution to construct a confidence interval— or rather we could if we knew π; but if we knew π, we would not be trying to estimate it. What we do is to estimate π by p throughout and assume that this estimate provides a good approximation.

Thus, in this case,

$$p \doteqdot N\left(p, \frac{p(1-p)}{n}\right)$$

and a 95% confidence interval for π is

$$p \pm 1.96\sqrt{\frac{p(1-p)}{n}}$$

$$= 0.4 \pm 1.96\sqrt{\frac{0.4 \times 0.6}{100}}$$

$$= 0.4 \pm 0.09602$$

$$= (0.30398, 049602)$$

Example. Of 180 selected applications for a Home Improvement Loan received by a Local Authority, 54 were successful. Construct a 90% confidence interval for the proportion of Home Improvement Loan applications which are successful.

$$p = \frac{54}{180} = 0.3$$

90% confidence interval

$$0.3 \pm 1.645 \sqrt{\frac{0.3 \times 0.7}{180}}$$

$$= 0.3 \pm 0.05619$$

$$= (0.2438, 0.3562)$$

Before moving on to the next section we ought to note that confidence intervals for a proportion involved two approximations. The first is the use of the Central Limit Theorem and the second is the use of p instead of π in the variance. This is often referred to as the Normal approximation to the Binomial. As a result of these approximations, the intervals are far more precise if p and π are close to 0.5 and if n, the sample size, is large.

12.4 Estimating a Variance

In most of the problems so far in this chapter, we have had to estimate the variance. We have used the sample variance as our estimate but we have never really investigated the question of how wrong or how inaccurate this estimate may be.

We can construct a confidence interval for σ^2 in much the same way as we did for μ and π. However, we need to introduce another distribution and use some statistical theory.

Suppose that we have a random sample of size n selected from a Normal population with variance σ^2. We calculate the estimated variance S^2. S^2 is unlikely to be exactly correct but we do know something about its variability for

$$\frac{(n-1)S^2}{\sigma^2} \sim \chi^2(n-1)$$

If we wish to construct a confidence interval, we pick the two endpoints of the interval by picking two points on the χ^2 distribution.

Thus, if we want a 90% confidence interval, we use the 5% point and the 95% point on the χ^2 distribution (with 90% between them). The confidence interval constructed will be

$$\frac{(n-1)S^2}{\chi^2_{0.05}(n-1)} < \sigma^2 < \frac{(n-1)S^2}{\chi^2_{0.95}(n-1)}$$

i.e. the 90% confidence interval for σ^2 is

$$\text{from} \quad \frac{(n-1)S^2}{\chi^2_{0.05}(n-1)} \quad \text{to} \quad \frac{(n-1)S^2}{\chi^2_{0.95}(n-1)}$$

Note that the χ^2 distribution is not a symmetric distribution and, therefore, the confidence interval we obtain will not be symmetric about the sample variance, S^2.

Suppose that we are trying to conduct a study into the salaries of newly-qualified chartered surveyors. In a sample of size 9, we have a variance of salaries of 160,000 (i.e. a standard deviation of £400). We wish to estimate the variance of salaries and so we construct a confidence interval. Suppose that we wish a 95% confidence interval for σ^2.

We have nine observations so we have eight degrees of freedom. We wish a 95% confidence interval so we use the 2.5% and 97.5% points of the χ^2 distribution (i.e. 95% between them).

Thus, we get an interval which goes from

$$\frac{(9-1)S^2}{\chi^2_{0.025}(9-1)} \quad \text{to} \quad \frac{(9-1)S^2}{\chi^2_{0.975}(9-1)}$$

$S^2 = 160,000$ and so the interval is from

$$\frac{8 \times 160,000}{17.535} \quad \text{to} \quad \frac{8 \times 160,000}{2.18}$$

i.e. $\quad$ (72,996.86, 587,155.96).

To construct a confidence interval for the standard deviation, we must construct a confidence interval for the variance and then take the square root of the end-points.

Thus, in this case, we had a sample standard deviation of 400 and we get a 95% confidence interval for σ of (£270.18, £766.26).

Suppose in this case, we collect an additional fifteen observations and we now have a sample variance of 100,000. Can we construct an 80% confidence interval for σ?

$$n = 24$$

80% confidence interval for σ^2 is

$$\frac{23 \times 100,000}{\chi^2_{0.1}(23)} < \sigma^2 < \frac{23 \times 100,000}{\chi^2_{0.9}(23)}$$

i.e. an 80% CI for σ^2 is

$$\left(\frac{23 \times 100,000}{32.007}, \frac{23 \times 100,000}{14.848} \right)$$

$$= (71,854.28, 154,903.01)$$

and so an 80% CI for σ is

$$(£268.07, £393.58)$$

12.5 Estimating Differences

12.5.1 Introduction

In this section, we shall be examining the problem of estimating the difference between two groups or between two populations. Often we are not interested in absolute values or prices but in comparisons and the four situations mentioned here all fall into that category. (However, this is an example of the Behrens–Fisher problem, see Section 13.7.2.)

12.5.2 Differences Between Two Population Means with the Same Variance

In this example, we have samples from two populations of size n_1 and n_2 respectively. The sample means from these samples are denoted $\bar{X}_1$ and $\bar{X}_2$ and the samples are assumed to be large enough that the CLT applies. The means of the two populations are denoted μ_1 and μ_2 and there is the assumption or the previous knowledge that these two populations have the same variance. However, we shall denote the sample standard deviation by S_1 and S_2, respectively.

Consider the following example. A medical researcher is investigating the difference in height between the two genders for 18 year olds. The researcher has the following data:

Males: $n_1 = 36$ $\bar{X} = 1.76\,\text{m}$ $S_1 = 0.1\,\text{m}$
Females: $n_2 = 39$ $\bar{X} = 1.61\,\text{m}$ $S_2 = 0.08\,\text{m}$

The researcher is interested in the differences and so we construct a confidence interval for the difference in population means, i.e. for $\mu_1 - \mu_2$.

Now,
$$\bar{X}_1 \doteqdot N(\mu_1, \sigma_1^2/n_1)$$
$$\bar{X}_2 \doteqdot N(\mu_2, \sigma_2^2/n_2)$$

and
$$\bar{X}_1 - \bar{X}_2 \doteqdot N\left(\mu_1 - \mu_2, \frac{\sigma_1^2}{n_1} + \frac{\sigma_2^2}{n_2} \right)$$

If we replace $\bar{X}_1 - \bar{X}_2$ by $\bar{W}$, $\mu_1 - \mu_2$ by μ_w and $(\sigma_1^2/n_1) + (\sigma_2^2/n_2)$ by $\sigma^2/f(n)$, then we have $\bar{W} \sim N(\mu_w, \sigma^2/f(n))$ and we can construct a confidence interval for μ_w in just the same way as we did for μ in section 12.2

As regards estimating σ^2 here, we have assumed that the variances are equal for the two populations and so we use a "pooled" estimate, denoted S^2p.

$$S^2p = \frac{(n_1 - 1)S_1^2 + (n_2 - 1)S_2^2}{n_1 + n_2 - 2}$$

$$= \frac{\Sigma(x_{12} - \bar{x}_1)^2 + \Sigma(x_{22} - \bar{x}_2)^2}{n_1 + n_2 - 2}$$

In this case, $S^2p = (0.34 + 0.2432)/73$

$$= 0.00799$$

For a 94% confidence interval, we have

$$\bar{X}_1 - \bar{X}_2 \pm 1.88 \sqrt{0.00799} \sqrt{\frac{1}{n_1} + \frac{1}{n_2}}$$

$$= 0.15 \pm 1.88 \sqrt{0.00799} \sqrt{\frac{1}{36} + \frac{1}{39}}$$

$$= 0.15 \pm 0.0388$$

$$= (0.1112, 0.1888)$$

(94% was chosen merely to try a different level of confidence.)

12.5.3 Differences Between Two Population Means with Unequal Variances

This situation is similar to the previous situation except that we no longer assume that the variances are equal. If we look at the previous problem, then, if we denote the two population variances by σ_1^2 and σ_2^2, respectively, we get

$$\bar{X}_1 - \bar{X}_2 \div N\left(\mu_1 - \mu_2, \frac{\sigma_1^2}{n_1} + \frac{\sigma_2^2}{n_2}\right)$$

The corresponding 94% confidence interval without the assumption of equal variances is

$$\bar{X}_1 - \bar{X}_2 \pm 1.88 \sqrt{\frac{S_1^2}{n_1} + \frac{S_2^2}{n_2}}$$

$$= 0.15 \pm 1.88 \sqrt{0.0002941 + 0.0001641}$$
$$= 0.15 \pm 0.04024$$
$$= (0.1098, 0.1902)$$

Thus, the interval which is constructed is usually but not always wider than the corresponding interval if we can assume equal variances.

12.5.4 Difference between Two Matched Populations

In this case we have two matched populations. We may have a set of houses and have their selling prices for the last two changes of ownership. Alternatively, we may have the income of a number of people where incomes are recorded for 1975 and for 1982. In psychological or biological tests, it is common to use identical twins and give them different courses of education or medication and examine the different effects.

In all of these cases, what is of paramount importance is the difference between the two members of a matched pair of observations, i.e. the difference in selling price for each house, the difference in income for each person, and the different levels of achievement between identical twins.

As a simple case we can look at data on 35 men who went on a six-month diet to lose weight. We have recorded the weights (in lb) before and after the six-month course.

Before:	193	241	214	173	185	191	168
	183	229	230	190	204	262	211
	195	221	252	235	218	241	186
	262	240	233	241	192	209	223
	216	186	265	233	244	210	217

After:	186	212	209	174	170	184	163
(in	191	203	224	178	185	231	201
same	184	209	223	217	192	223	187
order)	245	209	214	214	173	180	188
	223	172	213	226	219	214	183

As we are interested in weight loss, i.e. difference between weight before and weight after, we find the differences by subtraction and work with these.

Differences:	7	29	5	−1	15	7	5
(Before–After)	−8	26	6	12	19	31	10
	11	12	29	18	26	18	−1
	17	31	19	27	19	29	35
	−7	14	52	7	25	−4	34

If we now wish to estimate the difference between the mean weights before and after, this is the same as the mean of the differences and so we can take the difference figures and construct a confidence interval for the mean difference.

If we denote the ith difference by d_i, then

$$\Sigma d_i = 574 \quad \text{and} \quad \Sigma d_i^2 = 15{,}550$$

and so, to construct a confidence interval, we estimate the mean by

$$\frac{574}{35} = 16.4$$

and estimate the variance by

$$\frac{15{,}550 - (574)^2/35}{34}$$
$$= 180.48235$$

Thus, a 98% confidence interval is

$$16.4 \pm 2.33 \frac{\sqrt{180.48235}}{\sqrt{35}}$$
$$= (11.109\,\text{lb}, 21.691\,\text{lb})$$

12.5.5 Difference between Two Proportions

Of a sample of 220 commercial sites in South Yorkshire, 62 are fully occupied one year after completion of the site. Of a sample of 126 similar sites in the south-west of England, 49 are fully occupied one year after completion. Suppose that we are interested in the difference between the two area's occupation rates. In fact, whether we look at the proportion fully occupied or the proportion not fully occupied, we shall get equivalent answers.

We have to assume that our samples are independent random samples and, in general, we need samples of a reasonable size. We are trying to estimate $\pi_1 - \pi_2$ where π_1 is the proportion of all commercial sites in South Yorkshire which are fully occupied one year after completion and π_2 refers to the south-west of England.

As the samples are large and independent, if we estimate π_1 by p_1 and π_2 by p_2, and denote the sample sizes by n_1 and n_2, respectively,

$$p_1 - p_2 \doteq N(\pi_1 - \pi_2, V(p_1) + V(p_2))$$

where $V(p_1) = \dfrac{\pi_1(1 - \pi_1)}{n_1}$ and $V(p_2) = \dfrac{\pi_2(1 - \pi_2)}{n_2}$

Once again, we estimate π_1 by p_1 and π_2 by p_2 in the expression for the variance

and so $V(p_1)$ is estimated as $\dfrac{\dfrac{62}{220} \times \dfrac{158}{220}}{220} = 0.00092008$

and so $V(p_2)$ is estimated as $\dfrac{\dfrac{49}{126} \times \dfrac{77}{126}}{126} = 0.0018861$

To construct a 95% confidence interval for the difference between the two proportions, i.e. for $\pi_1 - \pi_2$, we have

$$p_1 - p_2 \pm 1.96 \sqrt{0.0009200 + 0.0018861}$$

$$= 0.281818 - 0.388888 \pm 1.96 \times 0.05297$$

$$= -0.10707 \pm 0.10382$$

$$= (-0.2109, -0.0033)$$

If we wish to look at the difference between the south-west England success rate and the South Yorkshire success rate, we merely multiply the end-points by -1 to get.

$$(0.0033, 0.2109)$$

12.6 Estimating a Correlation Coefficient

In Chapter 7, we saw that, if we have a discrete bivariate probability distribution, we can evaluate the correlation between two variables.

However, in many cases, the variables are continuous (or at least one is) or we do not have knowledge of the distribution. In these cases, we shall often be interested in estimating the correlation coefficient. Immediately we estimate the correlation coefficient, we have to confront the usual problem of constructing a confidence interval.

Unfortunately, the distribution of our estimate is not as simple as some of the distributions of estimates encountered earlier in the chapter. However, there does exist a good approximation for this distribution which we shall use.

If we take a simple random sample of size n from a large population and measure two variables on each selected unit and compute the product-moment correlation, r, between these two variables, then $\frac{1}{2}\left(\ln\left(\frac{1+r}{1-r}\right)\right)$ has an approximate Normal distribution with variance $\frac{1}{n-3}$, provided n is sufficiently large. Thus, if we denote $\frac{1}{2}\left(\ln\left(\frac{1+r}{1-r}\right)\right)$ by r^*, r^* has an approximate Normal distribution.

As an example, we have, in Table 12.1, data relating to a sample of 36 Building Societies in the UK. The data recorded are the number of borrowers and the number of branches. (Note that, strictly speaking, this is not a simple random sample as there are under two hundred Building Societies in the UK but we can use the data as an example.)

If we evaluate the correlation coefficient between these two sets of data, we get 0.9441. Thus, $r = 0.9441$. However, we need to construct a confidence interval.

r^* has an approximate Normal distribution and, if $r = 0.9441$, r^* equals 1.7745. Further, this statistic has a variance of $\frac{1}{33}$. Thus, our 95% confidence interval for r^* is $1.7745 \pm 1.96 \sqrt{\frac{1}{33}}$. This interval is from 1.4333 to 2.1157. Now this is the confidence interval for r^*. To get a confidence interval for the population correlation coefficient, we need to change the numbers back.

If

$$r^* = \frac{1}{2}\left(\ln\left(\frac{1+r}{1-r}\right)\right),$$

then

$$r = \frac{e^{2r^*} - 1}{e^{2r^*} + 1}.$$

Thus, we change back the endpoints of our interval for r^* and get a confidence interval for ρ of $(0.892, 0.971)$.

Table 12.1

	Number of borrowers	Number of creditors
Halifax	1,080,698	243,151
Nationwide	453,568	16,299
Bristol & West	74,810	19,222
London & South of England	40,398	5,701
Town & Country	33,743	2,430
Skipton	29,500	2,919
Newcastle	18,729	182
Dunfermline	14,492	967
City of London	7,396	1,095
Citizens Regency	4,930	241
Coventry Provident	5,963	184
Hinckley	5,623	67
Middleton	6,317	286
Mornington	3,673	330
Teachers'	2,687	255
City & Metropolitan	2,179	160
St. Pancras	1,441	94
Hemel Hempstead	2,046	1,149
Mercantile	2,862	164
Loughborough Permanent	2,456	1,002
Manchester	1,852	1,390
Peckham Mutual	1,543	129
Stafford Railway	1,745	22
Bath Investment and	1,429	20
Horsham	1,013	77
Herts & Essex	1,032	29
Tynemouth	1,285	35
Sheffield	921	61
Tynemouth Victoria	1,226	1,631
Cotswold	962	80
Bedford Crown	571	15
Clay Cross Benefit	675	0
Metrogas	467	0
Stockport Mersey	443	31
Chilterns	431	0
Musselburgh	246	13

Source: *Building Societies Association Bulletin, July 1982.*
Note: The name of the society, in each case, is formed by augmenting the words "Building Society" to the titles above.

In this case, the confidence interval is not very wide. This is due to the fact that we need quite a large sample for this approxima-tion to be valid and also because, with an estimate of 0.9441, we know that the true correlation must be quite high but can be no higher than 1.

If our estimate had been, for example, 0.2, then going through the same process will give a 95% confidence interval of $(-0.138, 0.496)$. As you can see, this interval is far wider and, in fact, includes a correlation of zero.

12.7 Comparison of Population Variances

Often we are interested in comparing two population variances to see if they can be treated as being equal or not.

Table 12.2

Valuation in isolation	Valuation with discussion
51,300	53,000
55,000	54,000
56,200	54,300
53,500	53,200
51,400	54,500
49,750	53,500
54,250	54,100
52,600	
$n = 8$	$n = 7$
$\bar{X} = 53,000$	$\bar{X} = 53,800$
$\Sigma x_i^2 - \dfrac{(\Sigma x_i)^2}{n} = 32,225,000$	$\Sigma x_i^2 - \dfrac{(\Sigma x_i)^2}{n} = 1,960,000$

Consider the data in Table 12.2. A residential property was requiring a valuation. Eight people valued the property in isolation while seven groups of people valued the property after a discussion.

The researcher who carried out this study considered it reasonable to assume that the two populations of valuations are approximately Normally distributed. Thus, the ratio of the estimated variances follows an F-distribution.

The F-distribution is indexed not by one but by two measures of degrees of freedom: one for the numerator and one for the denominator.

In our case, the estimate of the ratio is $\dfrac{32,225,000}{7} \Big/ \dfrac{1,960,000}{6}$ $= 14.0925$. Once again, an interval is a far more useful estimate and so we construct a 90% confidence interval for the ratio of the population variances σ_1^2 / σ_2^2.

The lower limit for this ratio is

$$F(0.95; n_2 - 1, n_1 - 1) \frac{s_1^2}{s_2^2}.$$

(Note that the degrees of freedom for the numerator is the same number as the divisor in the variance for the denominator.)

The upper limit is

$$F(0.05; n_2 - 1, n_1 - 1) \frac{s_1^2}{s_2^2}.$$

However, before we can calculate these limits, we must note that the F-distribution is tabulated for small values only. We can use the fact that

$$F(\alpha, n_2 - 1, n_1 - 1) = 1/F(1 - \alpha, n_1 - 1, n_2 - 1).$$

Accordingly, the lower limit for the ratio of the population variances is

$$[1/F(0.05, 7, 6)] \times 14.0925$$

and the upper limit is $F(0.05, 6, 7) \times 14.0925$.

The confidence interval, therefore, goes from

$$1/4.21 \times 14.0925 \quad \text{to} \quad 3.87 \times 14.0925$$

i.e. is $(3.3474, 54.5380)$.

As this interval does not include the value 1, we can probably conclude that group discussions appear to have an effect.

Formal tests can be carried out as will be seen in Section 13.11.

Note that, as with estimation of a standard deviation in Section 12.4, if we wish a confidence interval for the ratio of two population standard deviations, we must first construct an interval for the ratio of the variances and then take the square root of the limit of this interval.

12.8   Estimation—a Conclusion

Before we leave the topic of estimation, a few points ought to be made to put the preceding seven sections into perspective.

The first is that, throughout this chapter, we have taken the view of complete ignorance. We have assumed little knowledge about the situation and have tried to estimate certain parameters in order that we may gain some information and, therefore, some feeling for the range of the numbers with which we shall be dealing.

In some situations, however, we do have prior knowledge about the situation and we are trying to investigate if a change has occurred; for example, if inflation has increased prices or if government policy has had an effect. In these situations, we have an idea of a hypothesis and we are trying to see if our obervations will support or confirm this hypothesis or suggest that this hypothesis is implausible. This is the area of hypothesis testing which is dealt with in the next chapter.

Another important point is that we have been estimating the parameters of classical statistics: the mean, the proportion, the variance, etc. We have not tried to estimate the population median or mode or inter-quartile range. Unfortunately, the usefulness of these parameters is not yet widely enough accepted to justify dealing with estimating them—a process whose mathematics can be quite complex. Most statisticians would hope that eventually the median and the inter-quartile range replace the mean and the standard deviation as the accepted measures of location and spread. However, as that eventuality is a long way off, interval estimation of these parameters is omitted here.

Not all of this chapter need be understood exactly. The important points are as follows: the idea of using an interval estimate as a range of likely values, the general idea of construction of a confidence interval, and the problem caused by the fact that an interval of high confidence may be so wide as to be practically useless.

If these ideas are not clear in your mind, go back and read the relevant bits of the chapter—I suggest Section 12.1 and Section 12.2—before moving on.

Problems

1. A building society has recorded the number of enquiries they receive each day for loans for house purchases. For 53 days, the number of enquiries had a mean of 27.16 and a standard deviation of 9.46. Construct a 95% confidence interval for the mean daily number of enquiries for loans to purchase a house.
2. A sample of 28 shopping centres was selected and the number of shop units recorded. The mean number of units in this sample was 141 with a standard deviation of 36.3. Construct a 90% confidence interval for the mean number of shop units per shopping centre.
3. The costs for modernizing 42 selected homes of a given size and type had mean £3,781 and a standard deviation of £853. Construct a 99% confidence interval for the mean cost of modernizing one of this type of home.
4. Of a sample of 127 recent building contracts, 83 were complete

on time. Construct a 90% confidence interval for the proportion of recent building contracts which were completed on time.

5. Of 58 selected investment funds, 16 yielded a better return than a fund recommended by the *Financial Times*. Construct an interval of confidence 80% for the proportion of investment funds which fare better than the recommended fund.

6. Of a sample of 300 Scottish lawyers in 1984, 189 were in favour of allowing solicitors to advertise. Construct a 98% confidence interval for the proportion of solicitors who wished to allow advertising.

7. A sample of 11 solicitors were asked about their charges for a particular service. The variance of these charges was 10,816 (i.e. a standard deviation of £104). Construct a 95% confidence interval for the variance of these fees. Now, construct the corresponding confidence interval for the standard deviation of the fees.

8. Salaries of 25 selected people of about the same age and who perform the same job function are recorded. These salaries have a standard deviation of £310. Construct a 90% confidence interval for the standard deviation of the salaries of employees of the same age and function.

9. For 38 towns sampled from the West Midlands, the mean concentration of smoke and the number of approved home improvement loans were recorded. Their correlation was −0.601. Construct a 95% confidence interval for the correlation between these two variables for towns in the West Midlands.

10. For two dates, one month apart, in 1985, the price of 28 shares on the Stock Exchange was recorded. The correlation between these data was 0.27. Construct a 90% confidence interval for the correlation between the prices of shares on dates one month apart.

11. Two groups of workers are sampled as part of an investigation into how much overtime people work. From group A, 47 workers were selected who, in the period under study, worked a mean of 6.1 hours of overtime with a standard deviation of 2.2 hours. The 66 workers from group B had a mean of 4.3 hours with a standard deviation of 1.4 hours. Construct a 90% confidence interval for the difference between the mean amounts of overtime for the two groups.

12. To provide a rough estimate of national rates increases, a sample of 600 homes was selected in 1983. These homes had a mean rates bill of £426 with a standard deviation of £71. A further sample of 850 homes was selected in 1984. These homes had a mean rates of bill of £459 with a standard deviation

of £73.60. Construct a 95% confidence interval for the mean increase in rates. Would your interval construction have differed if we had available the rates bill for the same 600 houses for 1983 and 1984.

13. One hundred employees of a large surveying firm were sampled. Forty-one of these employees had a degree and/or a professional qualification. Of 120 employees of a competing firm, 43 have a degree and/or a professional qualification. Construct an 80% confidence interval for the difference between the proportion of employees in these organizations who hold a degree or a professional qualification.

14. The data in question 4 refer to building contracts in an area of Wales. In an area in the south of England, a sample of 210 building contracts had 177 completed on time. Construct a 98% confidence interval for the difference between these two areas in the proportion of their recent building contracts which are completed on time.

15. Two procedures for measuring and evaluating angles of elevation are being compared. A sample of eight students using procedure A had a standard deviation of 1.13° while a sample of nine students using procedure B had a standard deviation of 1.84°. Construct a 90% confidence interval for the ratio of the variances of angles of elevation reported by these students.

16. The group of people under study in question 8 are all male. We sample seven women from about the same age and same job functions as the men and their salaries have a standard deviation of £481. Construct a 95% confidence interval for the ratio of the standard deviations of the men's salaries to the women's salaries.

CHAPTER 13

Hypothesis Testing

13.1 Introduction

In this chapter, we shall be dealing with hypothesis tests—the fundamental statistical approach to making decisions. To try to understand different types of hypothesis tests, three examples will be discussed. After this, specific terms will be introduced and then some specific statistics examples will be covered in detail.

13.2 Hypothesis Tests—Three Situations

The first example is of the English court system. There is a defendant who, in fact, is either innocent or guilty. A trial is held, the result of which is that the defendant is either convicted or acquitted. The trial itself can be considered to be a hypothesis test. We have two competing hypotheses—innocence and guilt—and the hypothesis test is the trial itself. We take the action to acquit if it looks as if the defendant is innocent and we take the action to convict if it looks as if the defendant is guilty. However, conviction and guilt are not the same things; neither are acquittal and innocence. There may be some incorrect decisions and, unless we have perfect information, these are unavoidable. However, a fair trial will minimize the frequency of errors. (Readers may wish to note that, in the Scottish court system, there is an intermediate verdict which can be returned. This verdict is "not proven" and is interpreted as meaning that a decision could be made were full information available at the time).

The second example concerns a stock-broker. One of her clients wishes her opinion on whether to buy shares in Cashmore plc. Either the investment would be a sound one or it would be unwise. If we could see into the future, then we would be able to tell but we cannot. Unfortunately, the broker has to make a decision now and this decision may prove to be either correct or incorrect. Once again, we have two competing hypotheses—that the investment would be profitable and that the investment would lose money. The broker performs a hypothesis test. However, this test is not a formal test as in a court trial. She analyses the market conditions, company history, client's tax situation, etc., and from this, she reaches a decision and advises the client to buy or not to buy. Once again, she may offer incorrect advice but, hopefully, experienced brokers will make errors of judgement less frequently than brokers with less experience.

The third example is of a young couple looking for a new house. The first thing that they do is to go to an estate agency and look at the descriptions and photographs of several properties. On the basis of this information, they make a decision about whether to view the property. They do consider two competing hypotheses but this is not usually done consciously. The two hypotheses are that the property will be of ultimate interest and that the property will not be of ultimate interest. The trial is the investigation of the information provided by the estate agent and the appropriate actions are, respectively, to view the property or not to view. Once again, incorrect decisions can be made and can be costly. If the couple view a property which is not of interest to them, time is wasted. On the other hand, not viewing a property in which they would be interested constitutes a lost opportunity; this is something to which it is difficult to attach a cost but which is an important concept in Economics.

13.3 Parameters of Hypothesis Tests

Following on from the examples above, we can define some terms which will be important in the discussion which follows.

We shall be considering two competing hypotheses. These are called the *null hypothesis* and the *alternative hypothesis*. Generally, these will be denoted H_0 and H_1, respectively. To decide which is which, we generally designate the null hypothesis as the status quo or the default state and the alternative hypothesis as the hypothesis which we shall decide for if there is evidence of something different, i.e. what we are trying to prove.

Thus, in the three examples, the null hypotheses are "innocent", "poor investment" and "not worth viewing". The first of these stems from the "innocent until proven guilty" principle. The second is due to the fact that most clients will not invest until advised to do so, i.e. we would conclude that it is a poor investment unless we are fairly sure that it is a sound one. The third of these null hypotheses is due to the fact that, in general, people will not view a property unless they are quite keenly interested in it.

The corresponding alternative hypotheses are "guilty", "sound investment", and "worth viewing".

In a hypothesis test, we are testing the null hypothesis against the alternative hypothesis. We do this by assuming, temporarily, that H_0 is true and investigating whether, given the available evidence, it appears that H_1 is significantly more likely to be true than H_0. How much more likely it needs to be will be discussed below. If we decide that H_1 is far more likely, then we say that we reject H_0 in favour of H_1. If not, then we say that we fail to reject H_0. (This may seem strange terminology but, for several good reasons, we never accept H_0.) In carrying out this test, we must always remember that it is not known whether H_0 or H_1 is true.

If H_0 is true, then we can either make a correct decision—not to reject H_0—or an incorrect decision—to reject H_0. This latter decision is a mistaken decision and is called a *Type 1 error*.

On the other hand, if H_1 is true, then we can either make a correct decision—to reject H_0 and accept H_1—or an incorrect decision—to fail to reject H_0. This latter decision is a mistaken decision and is called a *Type 2 error*.

Thus, in the court system, convicting an innocent person is an example of a Type 1 error while acquitting a guilty person is an example of a Type 2 error. If we know something about the jury and the probabilities of each of them deciding one way or the other, we can calculate the probabilities of these two types of errors happening.

The probability of Type 1 error is denoted α and

$$\alpha = P(\text{reject } H_0 / H_0 \text{ is true})$$

The probability of Type 2 error is denoted β and

$$\beta = P(\text{fail to reject } H_1 / H_1 \text{ is true}).$$

We can see, and we shall see again later, that these two probabilities work against each other. If we wish to minimize the probability of convicting an innocent person, this is done at the expense of increasing the frequency with which guilty people are acquitted. Indeed, the only way to have the probability of convicting an innocent man to equal zero is to acquit everyone—not to have a trial at all. Thus, one of the costs of justice is to have a non-zero probability of convicting innocent parties.

Similarly, a stock-broker may wish not to recommend too many bad investments and be cautious. However, such caution will cause a number of good investment opportunities to be missed. On the other hand, if she wishes to cut down on the number of good investments missed, i.e. to be adventurous, this will entail more poor investments being recommended.

Two other measures are important here. As was stated earlier, the purpose is to examine the null hypothesis and, if there is sufficient evidence, to reject it and accept the alternative hypothesis. We can define the power of the test to be the probability that H_1 is accepted when, in fact, it is true. The power of a test is, therefore, equal to $P(\text{reject } H_0/H_1 \text{ is true})$. This equals $1 - \beta$.

Also, it must be realized that α and β vary as we change the test. Thus, what is generally done is to define the significance level of a test as $P(H_0 \text{ is rejected}/H_0 \text{ is true})$ and to fix this whenever we carry out a test. This probability equals α and for each test, we fix this significance level—which is usually quoted as a percentage—taking into account the gravity of the different types of errors and the probabilities of these types of errors occurring.

Thus, in the court system, we want a very low significance level to guarantee a very small proportion of convicted innocent parties. However, viewing a property which is of no interest is far less costly an error. In fact, it may be that, in this case, high power is desired so that almost all the properties which might be of interest (and, as an unfortunate result, a substantial number of undesirable properties) are viewed.

In statistical hypothesis tests, we derive what is called a test statistic which we shall denote T. In the court system, T is merely the number of votes of "guilty" from the jury. We then construct a simple decision rule of the form "If T is greater than (or less than) some value k, reject H_0, otherwise do not reject H_0". In the court case, the decision rule could be "Reject H_0 if T is greater than 8 i.e. greater than or equal to 9, otherwise do not reject H_0". The set of values of T which lead to rejection of the null hypothesis is called the critical region. Here the critical region is the set of values (9, 10, 11, 12). (Note that it is always possible to construct a test statistic T in such a way that the decision rule is of such a simple form. This is due to a basic theorem in statistics which is called the Neyman-Pearson Fundamental Lemma).

Clearly, if there is a full and fair trial, then there is a probability that each of the jurors will vote guilty. Thus, we could evaluate the probability that a guilty person is convicted or the probability that an innocent person is convicted etc. However, the problem is often that of deriving the probability distribution for the test statistic T.

The case of the trial is what is termed a one-tailed test. In this case, the alternative hypothesis appears on only one side of the null hypothesis and the critical region is in one piece. Sometimes we have to carry out a two-tailed test. Here we are looking for evidence on either side of the null hypothesis and generally the critical region will be in two pieces. For example, in testing if the mean heights of seven-year-old boys and girls are equal, we will look at the difference in mean height from a sample and test if it is close to zero. A large positive difference or a large negative difference will allow us to conclude that there is sufficient evidence of a difference.

We can now consider a statistical hypothesis test as being made up of seven stages.

(1) Clearly define the null hypothesis and the alternative hypothesis.
(2) Decide if the test is one-tailed or two-tailed.
(3) Choose an appropriate significance level.
(4) Select a test statistic T.
(5) Find the distribution of T when H_0 is true.
(6) Find the critical region and, if possible, the power of the test.
(7) Observe the data and reject H_0 if T falls in the critical region.

13.4 Hypothesis Tests for the Mean

Example 1. The mean salaries for employees in certain functions in Building Societies is £8,120. However, the New Town Building Society, which started business only a few years ago, has recently been on a drive to merge with other societies and to expand. As a result, there has been much interest in the salaries of their employees to see if the mean salary for employees in the same functions is different from £8,120.

We can set this up as a hypothesis test.

(1) H_0: Mean salary is £8,120
    H_1: Mean salary is not £8,120
(2) Let $\alpha = 0.05$, i.e. a 5%—significance level.
(3) Two-tailed test. Hence, we have $\alpha/2$ in each tail.
(4) Test statistics T is the mean salary from a sample of employees of the New Town Building Society.
(5) Critical region is a set of values t such that
    $P(T$ is in critical region$/H_0$ is true$) = 0.05$

If we take a large enough sample then, under H_0, by the CLT

$$T \doteqdot N\left(8,120, \frac{\sigma^2}{n}\right)$$

We can decide on a sample size n but we shall need to estimate σ^2 from the sample.

(6) We can construct a critical region of

$$\left(0, 8{,}120 - 1.96\,\frac{\sigma}{\sqrt{n}}\right), \text{ and } \left(8{,}120 + 1.96\,\frac{\sigma}{\sqrt{n}}, \infty\right),$$

i.e. the critical region is the union of these two intervals. If T, the sample mean, falls in this region i.e. in either interval, then we reject H_0. (Note that the 1.96 arises from a probability of 0.025 of being exceeded with a Normal distribution).

We can only find the power of a test against specific alternative hypotheses. We have here a general alternative hypothesis—in statistical terms, this is a composite alternative hypothesis—but the power of this test will be discussed in the next section.

(7) In a sample of 42 employees, there is a mean salary of £8,491 with a standard deviation of £872. If we use this sample standard deviation to estimate the population standard deviation, then the citical region is composed of the two intervals (0, £7,856.28) and (£8,383.72, ∞).

As our test statistic T falls in the critical region, we reject H_0. What in fact we are concluding is that the evidence collected and the calculated sample mean are so far from the null hypothesis that we are forced to view this null hypothesis as being highly improbable.

If, in this case, we suspected that the New Town Building Society was paying higher salaries then the differences would have been as follows. Firstly, the alternative hypothesis would have been that the mean salary was greater than £8,120. Secondly we would have a one-tailed test. The critical region would have had all the significance on one side of the null hypothesis and so would be the range

of values greater than $8{,}120 + 1.645\,\frac{\sigma}{\sqrt{n}}$ i.e. greater than £8,341.34.

As a result, while we still reject H_0, we have a different conclusion. In the two-tailed case, we can only conclude that the mean salary is not £8,120. Now, we can conclude that the mean salary is greater than £8,120. We should always decide beforehand what the alternative hypothesis is to be.

Example 2. An estate agent with 22 offices has recently begun to advertise in local newspapers. He wishes to ascertain the effect the advertising has had and is unsure about whether business will be increased by publicity or business will be hurt by people associating advertising with cheapness and unprofessionalism.

One measure of business is the number of enquiries that the offices have from potential clients. It is known from past records that the mean number of enquiries per office in a week in the same period of the year is 160.

The numbers of enquiries in an office has been noted for 2 randomly selected days in each of the 22 offices. Test if the advertising has had an effect.

Firstly, we can define the hypotheses. The null hypothesis is that advertising is of no beneficial effect. The alternative hypothesis is that advertising is beneficial. In this case, the hypotheses could be switched but, given the problem, the estate agent is likely to discontinue advertising unless it can be proved that it is of significant benefit.

Let $\alpha = 0.1$ i.e. we have a 10% significance level.

Clearly, it is a one-tailed test.

The test statistic T is the number of enquiries on the 44 selected days. These 44 observations are not independent—especially as there are pairs of observations from each of 22 offices. However, as a very rough approximation, we can invoke the Central Limit Theorem and assume that the mean of these 44 observations has an approximate Normal distribution. (Another reason why the Central Limit Theorem is problematic is that we do not have a random sample—it is a stratified random sample with a restriction of 2 per office). Despite all these problems we can proceed under the assumption of Normality and, if the result of the test is a close one, then we can examine the sensitivity of the test to the accuracy and validity of these assumptions.

T is then the mean number enquiries for the 44 days. If we know the previous variance and the variance for the sample, we would first have to test for the equality of the variances. However, let us use the variance of the observations.

The critical region is the set of values t for which $(T \geqslant t/H_0$ is true) $= 0.1$.

Suppose that we had a sample variance of 812. Thus, we want a set of values t such that $P(T \geqslant t/T \sim N(160, 812/n))$. This is the set of values of t greater than or equal to

$$167.07 \left(= 160 + 1.645 \times \sqrt{\frac{812}{n}} \right).$$

The power of the test depends on the particular value of the composite alternative we consider. Thus, we can evaluate the power for a mean of 163 enquiries or 167 enquiries or 173.6 enquiries. In fact, what is often done is to derive a mathematical function for the power by considering any values for the alternative hypothesis, calculating the power, and draw a graph.

The total number of enquiries was 7,387 which gives a mean of 167.886 enquiries. Thus, we reject H_0 and accept the alternative hypothesis that the advertising is proving beneficial. Subsequent research could look at a measure of the benefit and examine if it is greater than the cost of the advertising.

13.5 An Investigation into the Power of a Test

If we return to the first example of the previous section, we can look at the power of the test. You will recall that, in the court system example, the power of the test is the proportion of guilty people actually convicted and, in the stoke-broker example, the power of the test is the proportion of good investments which are recommended.

In the salary example, the power is the (proportion or) probability that the conclusion will be that the salary is not £8,120 when, in fact, the salary is different from £8,120.

If we take the standard deviation to be £872 and the sample size to be 42 then the power is the probability that the test statistic falls in the critical region if H_1 is true. The critical region is the combination of the intervals (0, £7,856.28) and (£8,383.72, ∞). However, we cannot evaluate the power unless we have a specific alternative hypothesis.

Suppose, for example, that the Building Society is paying a mean salary of £8,200. What is the probability of H_0 being rejected?

This equals

$P(0 < T < 7,856.28) + P(T > 8,383.72)$ if T has the approximate distribution $N(8,200, (872)^2/42)$.

This is a fairly basic Normal probability problem and the answer is $0.00523 + 0.0853$ which equals 0.0905 approximately.

If the alternative hypothesis is further away from the null hypothesis then the two will be more easily distinguishable. Suppose H_1 is that the mean salary is £8,300. Then, the power is 0.269.

If the mean under H_1 is £8,400, the power is 0.548 and if the mean under H_1 is £8,500, the power is 0.805.

Thus, for each specific value that we may care to consider as an alternative hypothesis we get a different value for the power of the test. The power will also change if we change the significance level of the test or the sample size.

For each hypothesis test, we can look at the whole range of alternative hypotheses, evaluate the power and plot them in a graph. These points will make up what is known as a power curve.

Test 1   Sample size 42. Significance level 5%
Reject H_0 if sample mean is in
(0, 7,856.28) or (8,383.72, ∞).
For H_1: mean = M
$T \div N(M, 18,104.38)$

We can plot the power curve for this test noting that the power at H_1: mean = £8,120.01 must be very close to 0.05.

Test 2   Sample size 70. Significance level 5%
Here, assume standard deviation of population is £872. The critical region, however, will alter and, for the same significance level, will be larger as we have more information.
Reject H_0 if sample mean is in
(0, £7,915.72) or (£8,324.28, ∞)

Once again, we can plot the power curve for this test.

Test 3   Sample size 42. Significance level 10%
The critical region will be larger than for test 1 as a higher significance level allows a higher probability of false rejections of H_0.
Reject H_0 if sample mean is in
(0, £7,898.66) or (£8,341.34, ∞)

This power curve appears with the other two below.

What would an ideal power curve look like? Well, we would want very low power very near to H_0 but the power would quickly shoot up to 1 as soon as we moved from H_0.

Is this shape of power curve attainable? As can be seen, increasing the sample size is a move in the right direction. In general, however, the ideal shape can only be attained by sampling all members of a population thereby collecting all the relevant information. Thus, the sample mean would be the population mean and the test would be 100% accurate.

(Note that power curves can be plotted for one-tailed tests. They will look similar in shape to the curves above on the side of H_0 where H_1 is to be found and will tend to 0 on the other side of the null hypothesis).

13.6   Hypothesis Tests for a Proportion

A nationwide driving school has a 28% pass rate for their students. However, one particular regional office last year presented 400

Fig 13.1

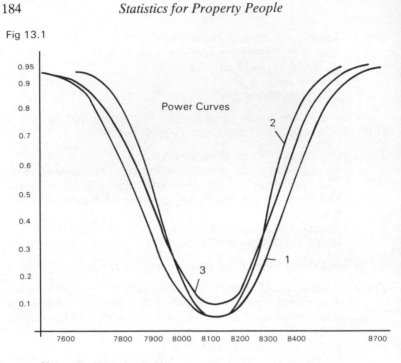

Power Curves

Alternative Hypothesis: $H_1 : \mu \quad =$

students for the driving test and 129 of them passed. Is this significantly better than the national success rate?

H_0: This region is doing as well as the rest of the nation
H_1: This region has a better success rate than the rest of the nation

If we denote the true success rate of this region by π, then what we have is

$$H_0: \ \pi = 0.28$$
$$H_1: \ \pi > 0.28$$

Let us carry out a 6% significance test which is, of course, a one-tailed test.

If we denote the observed success rate by p, then, under H_0,

$$p \div N\left(\pi, \frac{\pi(1-\pi)}{400}\right)$$

(Using the Normal approximation to the Binomial)

i.e. $p \sim N(0.28, 0.000504)$

Fig 13.2

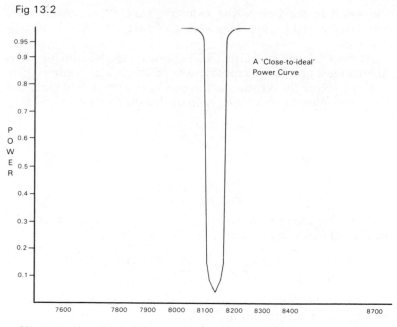

A 'Close-to-ideal'
Power Curve

Alternative Hypothesis: $H_1: \mu =$

Our test statistic T is the sample proportion p and we shall reject H_0 if T is in the critical region,

i.e. if T is $\geq 0.28 + \sqrt{0.000504}\ 1.55 = 0.315$

In fact, $T = 0.3225$ and so we reject H_0 and conclude that this region has a better than average success rate.

Alternatively, we can simplify the approach by rejecting H_0 if $P(T \geq t) \leq 0.06$. T, in the sample, is 0.3225 and so we can evaluate $P(T \geq t) = P(Z \geq 1.893)$ which equals 0.0294. We reach the same conclusion. However, while the arithmetic may be simpler for this one case, in general, we ought to formally set up a critical region so that we can issue instructions to other people to carry out a test and to report whether or not the value of T in their sample fell in the critical region.

13.7 Hypothesis Tests for a Difference of Means

A survey has recently been carried out by one of the employers' organizations. In thus study, 130 people were selected who were employed in the Computing industry. Half of the sample were employed in production rôles while half were employed in administration.

At the 4% level, test if the mean salaries in the two job functions are the same or whether employees in production earn more.

If we denote the mean salary in production by μ_p and the mean salary in administration by μ_A then the hypotheses are

$$H_0: \ \mu_p = \mu_A$$
$$H_1: \ \mu_p > \mu_A$$

or, more usually,

$$H_0: \ \mu_p - \mu_A = 0$$
$$H_1: \ \mu_p - \mu_A > 0$$

As we have seen in the last chapter, if we let T, our test statistic, be the difference in the sample means, then, under H_0,

$$T \doteqdot N\left(0, \frac{\sigma_p^2}{n_p} + \frac{\sigma_A^2}{n_A}\right)$$

Once again, the two population variances are unknown and so we must estimate them from the data.

Thus, $T = \bar{X}_p - \bar{X}_a$ and we shall reject H_0 if our test statistic falls in the critical region.

This critical region is from $0 + \sqrt{\dfrac{(2{,}031)^2}{65} + \dfrac{(1{,}843)^2}{65}} \times 1.75$

and so we reject H_0 if T is bigger than £595.30.

In our sample, T is £625 and so we reject H_0.

Once again, we could simplify the procedure by rejecting H_0 if $P(T \geq t) \leq 0.04$. In fact, $P(T \geq t) = P(Z \geq 1.84) = 0.0329$. Hence, the null hypothesis is rejected and we conclude that, at the 4% significance level, the mean salary in production is higher than the mean salary in administration.

13.7.1 The Behrens-Fisher Problem

The above problem of testing (or even estimating) the difference between two means when the population variances are unknown is known as the Behrens-Fisher problem and is one of the fundamental problems of Statistics. The approach we have followed is an approximate one and may be seriously in error if either of the samples is small or if either of the populations is radically different from a Normal distribution.

13.8 Hypothesis Tests for a Variance

In the area of investment, many problems arise because people's earnings fluctuate (and, therefore, so do their tax liabilities and upper tax rates). From past evidence, it is known that the variance of salaries of a group of the population is 9,000,000, i.e. the standard deviation is £3,000.

Recent wage negotiations have been held and an underlying premise was the maintenance of absolute differentials. If these are maintained, the variance of salaries ought to be unchanged.

A sample of 25 employees in this group was selected and they had a standard deviation of £3,622. Does this provide sufficient evidence to suggest that these differentials have not been maintained?

We can set this up as a formal hypothesis test.

$$H_0: \ \sigma^2 = 9,000,000$$
$$H_1: \ \sigma^2 \neq 9,000,000$$

I am presuming that we are interested in a shift in the variance in either direction.

Take α to be 0.05

Under H_0, $\dfrac{(n-1)S^2}{\sigma^2}$ has an approximate χ^2 distribution with $(n-1)$ degrees of freedom.

Thus, taking $\alpha/2 = 0.025$ at either end, we reject H_0

if $$\frac{(n-1)S^2}{\sigma^2} < \chi^2_{0.975}(24)$$

or if $$\frac{(n-1)S^2}{\sigma^2} > \chi^2_{0.025}(24)$$

i.e. if $$S^2 < \chi^2_{0.975}(24)\frac{\sigma^2}{(n-1)}$$

or if $$S^2 > \chi^2_{0.025}(24)\frac{\sigma^2}{(n-1)}$$

and $$\chi^2_{0.975}(24) = 12.401$$

$$\chi^2_{0.025}(24) = 39.364$$

Thus, we reject H_0 if

$$S^2 < 4,650,375$$

or if

$$S^2 > 14,761,500$$

S^2, in this case, is $(3,622)^2 = 13,118,884$. Thus, we fail to reject H_0.

If a one-tailed test had been carried out, the decision would have been

$$\text{Reject } H_0 \text{ if } S^2 < \chi^2_{0.45}(24)\frac{\sigma^2}{24} \qquad \text{for } H_1: \sigma^2 < 9,000,000$$

or

$$\text{Reject } H_0 \text{ if } S^2 > \chi^2_{0.05}(24)\frac{\sigma^2}{24} \qquad \text{for } H_1: \sigma^2 > 9,000,000$$

One-tailed tests of variance are quite common in many fields. This is because concern is usually about variances getting larger as a small variance means greater precision, and better information.

13.9   Hypothesis Tests for Equality of Variances

In continuing with the previous study, another group was sampled and 17 people were selected. The standard deviation of these salaries was £2,698. Does it appear that these two groups have the same variance?

If we let this group be denoted by a subscript 2, this test can be set up as previous tests.

$$H_0: \sigma_1^2 = \sigma_2^2$$
$$H_1: \sigma_1^2 \neq \sigma_2^2$$

Take $\alpha = 0.05$

Under H_0, $\dfrac{S_1^2}{S_2^2}$ has an F distribution.

This distribution is based on three main assumptions:

(i) The two samples were selected from populations with the same variance.
(ii) The samples were randomly chosen.
(iii) The populations are Normally distributed.

If assumptions (ii) and (iii) can be guaranteed, then a very large or very small value of this statistic can be used as evidence that the variances may not be the same.

As we are looking for either very small or very large numbers to reject H_0, the problem will be simpler by taking the larger estimated variance and dividing it by the smaller estimated variance.

If we take $\alpha = 0.05$, we can look up 0.025 in the tables as we have, in effect, folded the distribution over by having the larger estimate as the numerator.

Working out degrees of freedom is not a simple matter; especially as there are two of them. The first degree of freedom is the sample size of the estimate which is the numerator minus one. The other degree freedom is one less than the size of the sample which contributes to the denominator.

In this case,

$$F = \frac{(3,622)^2}{(2,780)^2}$$

and the degrees of freedom are (24, 16).

Looking at the table of the F distribution shows that the critical region is $(2.63, \infty)$. More explicitly, this hypothesis test says that, at the 5% level of significance, if the larger variance estimate is more than 2.63 times the smaller variance estimate, we must conclude that the population variances are not equal.

$$F = \frac{(3,622)^2}{(2,698)^2} = 1.802$$

Hence, we fail to reject H_0, i.e. we do not have sufficient evidence to conclude that the population variances are unequal.

This test is particularly useful when testing for the equality of two population means. As was stated earlier, testing for the equality of two population means is complicated by lack of knowledge about the variances. By carrying out a test of the equality of variances, it at least reveals if it looks as if the two populations have the same variance—and hence a pooled estimate is appropriate—or if the data sampled came from populations with differing variances.

13.10 Hypothesis Tests for Correlation

We have seen before how to calculate a correlation coefficient from a bivariate discrete distribution. However, in most cases, we have to estimate the correlation coefficient from a sample. In the previous chapter, we dealt with constructing a confidence interval for the population correlation coefficient. Now, we can discuss testing a hypothesis concerning its value.

If our null hypothesis is

$$H_0: \ p = 0$$

or $$\quad H_0: \ p \leqslant 0$$

or $$\quad H_0: \ p \geqslant 0$$

then we can construct a test using the t-distribution—an approximation used for constructing a confidence interval.

However, if the null hypothesis is anything else we need to employ a different method. This different method also applies for the above three cases and so we can recommend its use for all null values of p.

There is a hypothesis in the USA that, in cities which have expanded fairly recently, the cost of housing is greater the further from the city centre one is. In fact, it is hypothesized that the correlation between distance (in miles) from the city centre—as measured by the Court House—and price of a house is 0.38.

Now, this hypothesis has been shown to be valid for most areas but some research has recently been carried out in three cities: Fort Wayne, Indiana, Kansas City, Missouri, and Tulsa, Oklahoma. From these cities were selected 47 properties and the researchers were interested in testing if the previous correlation of 0.38 was maintained.

$$H_0: \ p = 0.38$$
$$H_1: \ p \neq 0.38$$

Our test statistic is r, the sample correlation coefficient. However, to derive a valid test, we use Fisher's Z-Transformation which we first met in §12.6.

This states that, under $H_0, \frac{1}{2} \ln \left(\frac{1+r}{1-r} \right)$ has an approximate Normal distribution with a mean of $\frac{1}{2} \ln \left(\frac{1+p}{1-p} \right)$ and a variance of $\frac{1}{n-3}$, where n is the size of the sample.

Thus, we must take our test statistic T to be $\frac{1}{2} \ln \left(\frac{1+r}{1-r} \right)$. Under H_0, this has a Normal distribution with mean $\frac{1}{2} \ln \left(\frac{1+p}{1-p} \right)$ with $p = 0.38$.

Thus, mean
$$= \tfrac{1}{2} \ln \left(\frac{1.38}{0.62} \right)$$
$$= \tfrac{1}{2} \ln (2.2258)$$
$$= \tfrac{1}{2} \times 0.80$$
$$= 0.40$$

and the variance $= \dfrac{1}{n-3} = \dfrac{1}{44}$

We have a two-tailed test with $\alpha = 0.1$ and so we reject H_0 if the test statistic fails to fall in the region

$$0.40 \pm 1.645 \times \sqrt{\frac{1}{44}}$$
$$= (0.152, 0.648)$$

Thus, the critical region for our test statistic is $(-\infty, 0.152)$ together with $(0.648, \infty)$.

Now, in the sample, the correlation coefficient was 0.211. However, this is not our test statistic. The test statistic is $\tfrac{1}{2} \ln \left(\dfrac{1+r}{1-r} \right)$ which equals 0.2142 which does not fall in the critical region and so we fail to reject H_0.

Note that if we had been testing H_0: $p = 0$, then the mean of the Normal distribution would have been $\tfrac{1}{2} \ln (1)$ which equals 0.

13.11 Hypothesis Testing—A Conclusion

The hypothesis tests we have examined are, to a great extent, merely the converse of the confidence intervals we developed in chapter 12. The only major change is that instead of estimating the unknown value of a parameter, we actually perform a test of a specific value of a parameter. However, if we prove that a particular value is very unlikely, we still need to estimate what value the parameter takes.

Hopefully, if chapter 12 was clear, then chapter 13 appears as only a slightly different framework. If that is not the case, then read §13.1 to §13.3 and then, for each parameter of interest, the corresponding sections of chapter 12 and chapter 13.

Problems

1. A building society's research department believes that the number of enquiries received each day for loans for house

purchase has a standard deviation of 9.46. Further, the mean number of enquiries per day is 14.16. A recent advertising campaign has affected the market and, over a 5-day period, 30 branches were selected to test the effectiveness of the campaign. Each selected branch was monitored for one day only and the mean of the recorded numbers of enquiries was 17.2. At the 5% level, test if the campaign had an effect.

2. Maintenance costs for the offices of a large company are being monitored. Prior to a reorganization aimed at reducing costs, the monthly costs (excluding wages) had a mean of £361.45 and a standard deviation of £33.64. Forty offices are closely monitored after the reorganization for a period of 3 months. The resultant 120 monthly costs had a mean of £347.92. At the 2% level, test if the reorganization had the desired effect. (Note that these 120 observations cannot, strictly speaking, be considered independent. However, for the sake of this example, we shall assume independence.)

   What is the power of this test if the reorganization actually reduces the mean cost by a) £5, b) £10, and c) £12.50?

3. Of a sample of 300 Scottish lawyers in 1984, 189 were in favour of allowing solicitors to advertise. At the 3% level, test if this proportion if significantly greater than 0.5?

4. In an effort to construct a simple index for share prices, a simple approach is to sample 75 share prices. The datum recorded for each share price in this simple approach is whether or not the price rose from the previous day's price. The statistics calculated is the total out of the 75 which rose. Set up a hypothesis test at the 5% level of whether the majority of share prices rose from the previous day. As a result of this test being set up, how many of the 75 would need to have risen for us to conclude that the majority of share prices have risen?

5. Economic theory suggests that cartels, on whatever scale, tend to cause distortion of prices. One feature of this distortion is that there will tend to be a great similarity in prices. In one market, the price of a service carried out by several competitors has a standard deviation of £16.40. However, government legislation has opened up this area to many more business. We take a sample of 27 businesses in this expanded market and ascertain their price for the service being examined. At the 10% level, set up a test of whether the cartel effect has been reduced by the government legislation. (Hint: set up the corresponding test for a variance.)

   Complete the test using the information that the standard deviation of the sampled prices was £18.92.

6. In manpower planning, age distribution of employees is very important. In one company, the standard deviation of the ages of employees is 7.2 years. However, in an external division of this company (which is administered separately), personnel practice has been different. 41 employees in this division are selected and their ages acertained. Set up a test, at the 1% level, of whether the standard deviation of ages in the external division is different from the rest of the company.

   Complete the test given the information that the standard deviation of the sample ages from the external division was 4.31 years.

7. Consider question 6 again. This company has a main competitor which for tax and legal reasons also has an external division which is operated separately from the rest of the company. These two external divisions are competing in a highly volatile market which requires non-standard hiring practices. A sample of 25 of this other company's external division's employees was selected. At the 5% level, test if the standard deviations of ages in these two external divisions are significantly different. The second sample had a standard deviation of 6.38 years.

8. In a sample of 265 homes, the correlation between the annual bill for rates and the annual bill for gas and/or electricity was found to be 0.275. At the 1% level, test if this is significantly greater than 0.16, a figure quoted in some earlier research.

9. As part of a study to gauge how good valuers are, 180 valuers were selected to value a domestic property. Two variables were measured: the number of years of valuation experience a valuer had and the relative error between the valuer's valuation and the selling price of the property.

   Only 161 of the properties were actually sold within the time span of the study and so only the data pertaining to these properties are included. (This may be a source of bias. Can you see why? Can you suggest what to do about it?)

   The correlation between these two variables was calculated. Set up a hypothesis test at the 5% level to test if this correlation is significantly different from zero.

   The correlation coefficient for the 161 properties was, in fact, -0.19. What do you conclude?

10. Consider the problem posed in question 11 of chapter 12. At the $2\frac{1}{2}$% level, test if we can conclude that the workers in Group A work, on average, more overtime than the workers in Group B.

PART SIX

Applied Problems

CHAPTER 14

Analysis of Time Series

14.1 Introduction

So far, in this text, most of the data with which we have dealt have been thought of as being collected at a moment in time or in a small interval of time. However, sometimes data are collected over a long period of time. In these cases, the data generally exhibit certain features which cause us to use a different method of analysis. If, for example, we collect quarterly figures of mortgage advances, we call the sequence of observations a time series and the specific approach or method of analysis is Time Series Analysis.

Other examples of time series are:

the price of petrol each Monday morning;
the monthly unemployment figures;
the number of cars sold each quarter;
the frequency of trains on a particular route each hour.

In this chapter, we shall use statistical models in an attempt to analyse time series. In particular, we shall attempt to identify, estimate, and isolate several components which together comprise the actual numbers which appear in the sequence of observations. Further, once we have some idea of the components of the time series, we can then use them for making policy decisions and/or for projecting the time series into the future. It must be noted that such projections into the future may be inaccurate for at least two reasons. Firstly, we are only estimating the components; we do not know their exact values. Secondly, in general, we shall be making an assumption of the stability of social, political, economic, and demographic conditions.

As an example, consider the number of housing starts, recorded quarterly, for the last twenty-five years. It is clear that, in general, the number of housing starts has increased. This is due to an increase in population and in demolition, greater mobility, etc. This general movement is called a *trend*. Most time series contain a trend (upward or downward) which results from the long-term influence of various socio-economic and other factors. However, the major characteristic of a trend is that it continues in one direction for a long period of time.

We must also consider the fact that the number of housing starts will be high in the spring, reach a peak in the summer, and then fall off in the autumn and winter. Thus, there is some type of *seasonal component*. Seasonal components are also evident in unemployment figures, ice cream sales, etc. In fact, in the last few years, the unemployment figures which have been quoted by the UK government are seasonally-adjusted, i.e. they have had the seasonal component estimated and removed.

Note that seasonal components need not correspond to the actual seasons. Neither need there be four of them. Indeed, use of the term "seasonal components" is loose terminology. If we have a monthly series, we could isolate twelve seasonal components, and if we have a daily time series, we could isolate seven seasonal components—corresponding to the seven days in the week.

Housing starts have also been subject to a second type of periodic movement which is more irregular and longer in duration than the seasonal components. These fluctuations can often be explained in terms of the economic state of the country which, of course, will be affected by changes in government, interest rates, building society policy, housing policy, etc. Economic theory suggests that these effects occur quite regularly and we call them *cyclical components*. An early recorded cycle is a seven-year cycle of wheat prices which is evident for most of the 19th century. In addition, in the United States of America, we would expect a strong four-year cycle—corresponding to the fact that there is a Presidential election every four years.

Finally, although we have identified three components, we will not have completely explained the number of housing starts that we actually observed. There is likely to be a difference between what we observe and what we would expect from the three components. This difference is called the *residual component*, and, in general terms, this is what cannot be accounted for by the trend, the seasonal components, or the cyclical component and is the result of random fluctuations and/or once-only occurrences. Of course, if it is possible to provide a fairly accurate estimate of the effect of a once-only occurrence, e.g. a currency devaluation or a change in the rate of tax relief, this estimate can and, indeed, should be incorporated as a separate component into the model.

14.2 The Basic Models

In what follows, we shall denote the actual observations in the time series by the letter A. The trend values shall be denoted by T, the seasonal components by S, the cyclical component by C, and the residual component by R. In addition, we shall try to analyse time series using two basic models: the Additive Model and the Multiplicative Model. Many different models are possible as are hybrids of models. However, these two are quite common, are relatively simple to use, and are also relatively accurate.

The Additive Model states that the actual observations in the time series arise from taking the sum of the individual components. Thus, in notational terms,

$$A = T + S + C + R$$

On the other hand, the Multiplicative Model states that the actual observations in the time series arise from taking the product of the individual components. In notational terms,

$$A = T \times S \times C \times R$$

14.2.1 Smoothing—Moving Averages

Most sets of data display fluctuations which occur at random. If possible, we would like to remove these. However, the best that we can usually do is to reduce the effect they have on the analysis. This is done by smoothing the data.

The most common method of smoothing the data is by calculating a *moving average*. This tends to reduce the effect of single fluctuations by spreading out the unevenness.

A moving average of a time series is obtained by taking each sequence of n successive observations and replacing them with the mean of the sequence. (The terminology "n-point moving average" is also used.) The number n denotes the term of the moving average. In most cases, n should make sense in light of the nature of the data. For example, if the data are recorded monthly, $n = 12$ would be sensible. For daily data relating to shop sales, $n = 6$ might be advisable. Sometimes, a suitable value of n can be suggested by a plot of the data.

As an illustration, we can look at the data in Table 14.1 below. The first column of data contains the number of telephone enquiries for a particular property which a Glasgow solicitor receives for a 20-day period, i.e. 4 weeks, Monday to Friday.

In the second column are the sums of successive sequences of five numbers. I use five as there are five working days in this solicitor's week. Thus, the first number is $6 + 31 + 26 + 20 + 11 = 94$. The second number is $31 + 26 + 20 + 11 + 9 = 97$, i.e. the sum of the second set of five numbers. The reasoning for the positioning of these successive sums will become clear soon.

For the third column of data, we divide the figures in the second column by 5 to get the mean. These mean figures are now centred, i.e. positioned according to the middle number which contributes to that mean. This third set of figures is the 5-term or 5-point moving average.

Table 14.1

Day	Number of Enquiries	5-term Moving Sum	5-term Moving Average	7-term Moving Sum	7-term Moving Average
1	6				
2	31				
3	26	94	18.8		
4	20	97	19.4	133	19.0
5	11	96	19.2	149	21.3
6	9	92	18.4	136	19.4
7	30	90	18.0	124	17.7
8	22	93	18.6	112	16.0
9	18	92	18.4	139	19.9
10	14	100	20.0	153	21.9
11	8	101	20.2	143	20.4
12	38	103	20.6	131	18.7
13	23	99	19.8	121	17.3
14	20	99	19.8	148	21.1
15	10	102	20.4	169	24.1
16	8	108	21.6	153	21.9
17	41	110	22.0	144	20.6
18	29	114	22.8		
19	22				
20	14				

You will see that the moving average does not exhibit the wild fluctuations caused by the effects of the days of the week. (Tuesday is "property day" in the Glasgow newspapers.)

In the last two columns are calculated a 7-term moving sum and a 7-term moving average. As can be seen from the last column, the fluctuations have been reduced but not by as much as with the more appropriate 5-term moving average.

If the term of the moving average is even, then the mean will not correspond to one of the original data points. The solution to this is in the next section.

14.3 Analysis with an Additive Model

The data to be analysed are the number of houses changing hands in a particular area, recorded quarterly for five years from 1978 to 1982. The quarters shall be denoted I, II, III, and IV.

	I	II	III	IV
1978	136	170	195	141
1979	159	178	204	163
1980	170	184	210	185
1981	178	201	214	190
1982	193	211	225	200

If these data are drawn on a graph, several points can be made. Firstly, there seems to be a general overall increase. Secondly, there seems to be a regular pattern around this general overall increase illustrating the effects of the quarters.

Fig 14.1

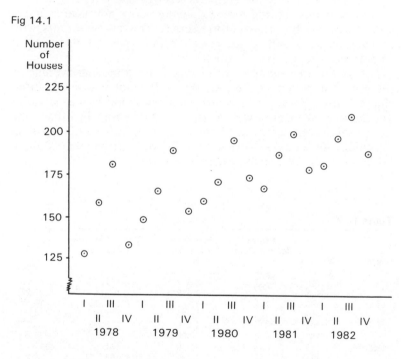

14.3.1   Additive Model—The Trend Component

The simplest way to represent this general increase or trend is by a straight line. Over a short period, a straight line rarely presents problems although, if, in fact, the trend is curved, e.g. increases by 3% per year, then a straight line will not be appropriate for a 20-year time series.

However, before trying to produce a straight line, let us get a suitable moving average of this time series. First of all, we get a 4-term moving sum. However, if we were to take the mean of these and place it at the centre of the four values, this does not correspond to one of the original values. To get around this, we add together successive pairs of sums and the mean of this is called a centred moving average.

Another way to think about this is as a weighted moving average. The first four-term moving sum is $T_1 + T_2 + T_3 + T_4$. The second sum is $T_2 + T_3 + T_4 + T_5$. Thus, the first figure in the moving average is $(T_1 + 2T_2 + 2T_3 + 2T_4 + T_5)/8$ i.e. we have a weighted 5-term moving average where the numbers receive weights 1, 2, 2, 2, 1.

Note that this centred moving average does not have as many data points as the original time series. This always occurs with moving averages and is the price we must pay for this type of smoothing.

Let us, for the moment, treat this centred moving average as the trend component of the time series. It is not an exact relationship but the values are increasing fairly steadily by just under 3 per quarter. We can now construct a set of data which have the trend component removed. This appears in Table 14.2 below in the column headed $A–T$—we subtract as we are using the Additive Model—and is called a detrended time series.

Table 14.2

			4-term Moving Sum	Sum of two Moving Sums	Centred Moving Average	A–T
1978	I	136				
	II	170				
	III	195	642	1,307	163.375	31.625
	IV	141	665	1,338	167.25	−26.25
1979	I	159	673	1,355	169.375	−10.375
	II	178	682	1,386	173.25	4.75
	III	204	704	1,419	177.375	26.625
	IV	163	715	1,436	179.5	−19.5
1980	I	170	721	1,448	181	−11
	II	184	727	1,476	184.5	− 0.5
	III	210	749	1,506	188.25	21.75
	IV	185	757	1,531	191.375	− 6.375
1981	I	178	774	1,552	194	−16
	II	201	778	1,561	195.125	− 5.875
	III	214	783	1,581	197.625	16.375
	IV	190	798	1,606	200.75	−10.75
1982	I	193	808	1,627	203.375	−10.375
	II	211	819	1,648	206	5
	III	225	829			
	IV	200				

14.3.2 Additive Model—Seasonal Components

If we look at the detrended series, we can see a fairly regular pattern of large and small numbers. We can use these figures to estimate the seasonal components.

Consider, first of all, the figures for the first quarter of each year. For 1978, we have no trend figure and so we have to omit this period from an estimation of seasonal components.

For 1979, the trend was 10.375 above the observed figure. For 1980, the trend was 11 above the observed figure while, for 1981 and 1982, the trend was 16 and 10.375, respectively, above the observed figures.

It appears that the first quarter figures are all substantially below what one would expect from only using the trend. Thus, we can estimate the seasonal component for the first quarter as the mean value for the first quarter of the corresponding figures in the $A-T$ column.

Estimated component for quarter I

$$= (-10.375 - 11 - 16 - 10.375)/4$$
$$= -11.9375$$

In a similar way, we can estimate the components for the other three quarters.

Estimated component for quarter II

$$=(4.75 - 0.5 + 5.875 + 5)/4$$
$$= 3.78125$$

Estimated component for quarter III

$$= (31.625 + 26.625 + 21.75 + 16.375)/4$$
$$= 24.09375$$

Estimated component for quarter IV

$$= (-26.25 - 19.5 - 6.375 - 10.75)/4$$
$$= -15.71875$$

We now have four estimated seasonal components. However, we must check that they make sense in our model. The most important check we do is to add up the seasonal components. In an Additive Model, they ought to sum to something close to zero. If this is not so, then it means that the seasonal components themselves have a sort of trend in them in that, over the space of each year, they either cause the figures to decrease or cause them to increase.

In this example, they sum to 0.21875. This is probably quite acceptable but to be precise and to assist in the rounding, I shall make the necessary adjustment.

If the four numbers add up to 0.21875, then subtracting $(0.21875)/4$ from each will cause them to add up to zero. Having done this, I rounded off the components to get $(-11.99, 3.73, 24.04, -15.77)$ which add up to 0.01. Thus, we are estimating that, on average, first quarter figures are 11.99 below the trend, second quarter figures are 3.73 above the trend, third quarter figures are 24.04 above the trend, and fourth quarter figures are 15.77 below the trend. Both because these components are estimates and because they are average effects means that there is no reason for them to be whole numbers.

We can now construct a new series which has the seasonal components removed as well as the trend. This is a detrended deseasonalized series and is denoted in Table 14.3 by $A-T-S$.

14.3.3　Additive Model—The Cyclical Component

Estimating the cyclical component is not an easy matter. What is often of importance is to be able to predict turning points in the economy, i.e. points at which a cycle begins to rise and points at which a cycle begins to fall. However, in many well-studied cases the length of the cycle, i.e. the time between successive peaks, is itself often varying quite markedly.

If we consider a series in isolation, it is nearly impossible to identify the cycles. Therefore, in most economic contexts, what is done is to look at several variables at a time and to examine the relationship among them. For example, in our case, we are looking at housing sales. If the economy is about to reach a high part of a cycle and then fall off, which variables might we expect to see falling before house sales? A decrease in the number of housing sales is likely to be caused by a cut in the availability of money to be borrowed and/or the fear or the reality of increased interest rates. Therefore, if the interest rate has been dropping steadily and so encouraging business in house sales and then stops dropping, some months later, the number of house sales will reach a peak and fall off as fears will grow of a rise in interest rates. In a similar way, fears of a motorway development will cause house sales to fall as will a fall in the general economy of the country or an increase in stamp duty and/or conveyancing fees.

In addition, the type of model which is often used for the cyclical component is not simple mathematically.

To accurately estimate the cyclical component also requires a lot of data, e.g. if a six-year cycle was suspected, we would probably need at least twenty years of data to be able, firstly, to verify that six years is about the correct length of the cycle and, secondly, to estimate the amplitude of the cycle, i.e. how high are the peaks and how low are the troughs.

For these reasons, we shall not, in this text, attempt to estimate the cyclical components. We should, however, be aware that for a long-term study such estimation is important.

14.3.4 Additive Model—Residual Component

The residual component is what is left after we have removed the effects of the trend, the seasonal components and, if estimated, the cyclical component.

The residual component is what has not yet been explained or accounted for and must be examined carefully. We often use the set of residual components as a form of diagnosis. What we are seeking is some pattern which will reveal, perhaps, the existence of some identifiable component which we have not yet introduced. Alternatively, we may see a pattern which suggests that the original model is not valid.

Another reason for examination of the residuals is to try to ascertain how close our component estimates may be and, more importantly, how accurate any predictions we make may be (*vid* §14.4).

In our example, if we ignore the cyclical component, then the column headed $A–T–S$ in Table 14.3 below contains the residual components. The types of pattern which may appear include the following.

(1) A group of residuals which are mostly negative followed by mostly positive residuals, or vice versa. This suggests a once-only shift in the economy, e.g. change in tax rate, devaluation.

(2) The difference between successive residuals becoming steadily larger/smaller, possibly in conjunction with (1). This suggests that a multiplicative model may be more appropriate.

(3) One residual wildly different from the rest. This suggests an arithmetic error or a recording error.

(4) A recognizable pattern of positive and negative residuals. This suggests that a strong cyclical component has not been identified or that the period of the moving average may be wrong.

Other patterns are, of course, possible. What we are looking for is a set of residual components which is composed of small numbers with no discernible pattern. Any pattern ought to be explained or, by introducing another component, removed.

As regards the set of residual components in Table 14.3 below they would appear to be in order. What they tell us is, for example,

Table 14.3

		A–T	S	A–T–S
1978	III	31.625	24.04	7.585
	IV	−26.25	−15.77	−10.48
1979	I	−10.375	−11.99	1.615
	II	4.75	3.73	1.02
	III	26.625	24.04	2.585
	IV	−19.5	−15.77	− 3.73
1980	I	−11	−11.99	0.99
	II	− 0.5	3.73	− 4.23
	III	21.75	24.04	2.29
	IV	− 6.375	−15.77	9.395
1981	I	−16	−11.99	− 4.01
	II	5.875	3.73	2.145
	III	16.375	24.04	− 7.665
	IV	−10.75	−15.77	5.02
1982	I	−10.375	−11.99	− 1.615
	II	5	3.73	1.27

that the actual figure for the third quarter of 1978 is 7.585 above what we have so far explained and that the actual figure for the fourth quarter of 1978 is 10.48 below what we have so far explained.

The five residuals whose absolute size is above 5 should be checked for arithmetic accuracy. There does not appear to be a pattern. Therefore, all that remains is to quantify the residuals. As with the seasonal components, the mean ought to be close to zero. The mean of the residuals is 0.0521875 which is probably close enough. We can estimate the variance of the residuals to be 27.52. Thus, we have the standard deviation of the residuals as 5.246 house sales.

14.4 Forecasting

While one main purpose of time series analysis is the identification and estimation of the various components, another main purpose is to allow prediction or forecasting of future figures.

To do this, we must be very careful about the assumptions we make, either explicitly or implicitly. Some of these assumptions will be discussed below.

Let us again consider the data on the number of houses changing hands. Suppose that we wish to forecast the figures for 1983. How do we do it?

14.4.1 A Trend Line

The first problem occurs with the trend. How does one extend the moving average, which is smoothed but not smooth, into the future? The simple answer is to use a straight line. There are many ways to fit a straight line. Four will be detailed in this text but only three of them in this chapter. The four which appear in this text are

(1) By eye
(2) By the Method of Semi-Averages
(3) Resistant Line
(4) Regression Line

14.4.2 Fitting a Trend Line by Eye

To fit a line by eye, one could plot the points—the values of the centred moving average in Table 14.2 above—on a graph and move around a piece of string or a transparent ruler until one has what appears to be a "good line". By good, I mean that the line appears to represent the points fairly well.

We cannot say that one line is correct and one is wrong because we do not know what the correct line is. However, what we do know from experience is as follows.

Firstly, the same person will not settle on the same line two days in a row. There may be slight differences or quite substantial differences. Which of two lines is someone to use? One cannot really take an average of two lines.

Secondly, the brain and the eye tend to place too much importance on points which seem to differ from the rest and not enough importance on those points which establish the trend.

Thirdly, if one person is going to produce different lines on different days, how are several people acting independently going to perform? They are unlikely to reach any sort of agreement on a line to use and, therefore, will not make any progress.

Fourthly, and finally, it is important to know that a line being used is accurate and to be able to quantify the accuracy. This is impossible to do if fitting a line by eye and, therefore, I would recommend fitting a line by eye only for use in a very rough way.

14.4.3 Fitting a Trend Line by Semi-Averages

This is a simple method which gives a fairly representative line. It relies on the fact that only two points are needed to specify completely a straight line. Therefore, what we do is to split the data into two groups. We then obtain one point which will be used to represent, in some way, the first group and one point which will be used to represent the second group. Then, we draw a line through these two points.

In the example on house sales, there are sixteen numbers in the centred moving average. We take the first eight as one group and the second eight as the second group. Now, if we number the original figures from 1 to 20, then the figure of 163.375 corresponds to the third quarter, 167.25 to the fourth quarter, 169.375 to the fifth quarter, and so on.

Quarter	Group 1 Centred Moving Average	Quarter	Group 2 Centred Moving Average
3	163.375	11	188.25
4	167.25	12	191.375
5	169.375	13	194
6	173.25	14	195.125
7	177.375	15	197.625
8	179.5	16	200.75
9	181	17	203.375
10	184.5	18	206

You will see from the graph below, what these data look like. We shall try to identify the two points to represent the groups by taking the means of each column above.

Thus, for the first group, the mean figure occurs at quarter number 6.5 and is 174.453125. (In this context of obtaining representative points, we can ignore the fact that there is no quarter 6.5). For the second group, the mean figure occurs at quarter number 14.5 and is 197.0625.

We now have our two points and we can draw a line between them. (The equation of a line appears in chapter 4). The equation of this line is approximately

$$Y = 156.085 + 2.826X$$

or

$$\text{Trend} = 156.85 + 2.826 \text{ (Number of quarter)}$$

Thus, the trend seems to be rising by about 2.826 per quarter. This corresponds with our estimate in §14.3.1 that the trend was rising "by just under 3 per quarter".

This method is not the most accurate mathematically but it does resolve the first three problems which occur with fitting a line by

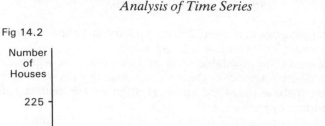

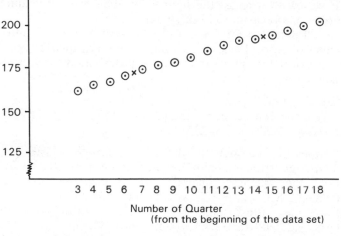

Fig 14.2

Number of Quarter
(from the beginning of the data set)

eye. As regards accuracy, this method is quite good and some simulation studies seem to support this.

14.4.4 Short-term and Long-term Forecasting

Forecasting of any sort whether it be weather forecasting, fortune telling or commodity price forecasting is hazardous. In our case, forecasting is hazardous for three reasons.

(1) It relies on the validity and completeness of the model.
(2) It relies on the accuracy of the estimates.
(3) It relies on the stability of the environment.

Any forecast relies on the model used being appropriate, valid and complete. If a significant component is omitted, the forecast values may be greatly in error.

Any forecast also relies on the fact that we have estimated several components. These estimates are almost certain to be in error but we can hopefully assume that the errors will be small and not troublesome.

Any forecast also relies on the environment or the economy being stable. For example, if the economy is presently growing at 2% p.a., in making a forecast, we must assume that the economy will continue to grow at roughly 2% p.a. at least up to and including the time for which the forecast is being made.

If the time of the available data and the time for which the forecast is desired are close, then we are dealing with short-term forecasting. If these times are far apart, then we are dealing with long-term forecasting.

If we try to forecast the figures for 1983, we are dealing with short-term forecasting and carry it out as follows.

Recall that $A = T + S + C + R$. We have omitted C, and R has a mean of zero. Thus, our estimate for each quarter of 1983 is the sum of the trend figure for that quarter and the corresponding seasonal component.

1st quarter 1983

This is the 21st quarter in the series so the estimated trend value is $156.085 + 2.826 \times 21 = 215.431$

The seasonal component for the first quarter is -11.99.

Thus, our estimated figure for the first quarter of 1983 is 203.441.

For the second quarter of 1983, the estimate is $156.085 + 2.826 \times 22 + 3.73$ which equals 221.987.

For the third quarter of 1983, the estimate is $156.085 + 2.826 \times 23 + 24.4$ which equals 245.483.

For the fourth quarter of 1983, the estimate is $156.085 + 2.826 \times 24 + 15.77$ which equals 208.139.

If one wishes, these figures can be rounded to the nearest whole number. However, more importantly, we ought to give some idea of how accurate these are. One way to do this is to quote the standard deviation of 5.246 house sales. Another way is to construct confidence intervals and these will be dealt with in the next chapter.

This short-term forecast relies quite heavily on the cyclical component being small enough that our omission of it does not cause serious errors. If this is true, then we can be fairly confident in the accuracy of the short-term forecast. However, in many cases, the cyclical component should not be omitted and/or the residuals are quite large. In these cases, short-term forecasting is not possible but long-term forecasting is. Here, we are mainly relying on the stability of the trend. If, for example, we wished an estimate for 1990 and could assume that the trend will still hold for that period of time, then our estimate for the third quarter of 1990 would be 324.611. This is quite a wide range but, when we consider that we are trying to predict eight years ahead, is relatively narrow. In effect, most of this estimate is contributed by the trend component which can, in certain situations, be assumed to be stable over very long periods of time.

Thus, a long-term forecast is often really a forecast of the trend as the seasonal components and cyclical components will be relatively small. (This is not true in the minority of cases where a trend which can be assumed to be stable over a long period of time is, in fact, a falling trend.) If the short-term fluctuations and seasonal components are secondary in importance to the long-term behaviour of the trend, then a long-term forecast is desirable.

14.5 Analysis with a Multiplicative Model

To ease matters, we shall analyse the same set of data but use a different model. The basic approach is the same except that we use a multiplicative model and so the actual components will have different interpretations.

14.5.1 *Multiplicative Model—The Trend Component*

Before getting a straight line for the trend, we ought to perform some smoothing. Therefore, we shall use as a starting point, the centred moving average of Table 14.2 above.

We must remove this trend but, as we are employing a multiplicative model, we do not subtract the trend from the actual data but divide the trend into the actual data. This is done in Table 14.4 below.

14.5.2 *Multiplicative Model—Seasonal Components*

The seasonal components in the multiplicative model can be evaluated in at least two ways: by an arithmetic mean or by a geometric mean.

(i) By arithmetic mean, we merely perform the same operation as was carried out for the additive model.

Thus for the quarter I the estimated component is

$$(0.9387 + 0.9392 + 0.9175 + 0.9490)/4 = 0.9361$$

For quarter II, the estimated component is

$$(1.0274 + 0.9973 + 1.0301 + 1.0243)/4 = 1.0198$$

For quarter III, the estimated component is

$$(1.1936 + 1.1501 + 1.1155 + 1.0829)/4 = 1.1355$$

For quarter IV, the estimated component is

$$(0.8430 + 0.9081 + 0.9667 + 0.9465)/4 = 0.9161$$

Table 14.4

		Centred Housing Average (T)	A/T (to 4 d.p.)	S	A/(T × S)
1978	I	136			
	II	170			
	III	195	163.375	1.1936	1.135	1.052
	IV	141	167.25	0.8430	0.916	0.920
1979	I	159	169.375	0.9387	0.936	1.003
	II	178	173.25	1.0274	1.019	1.008
	III	204	177.375	1.1501	1.135	1.013
	IV	163	179.5	0.9081	0.916	0.991
1980	I	170	181.0	0.9392	0.936	1.003
	II	184	184.5	0.9973	1.019	0.979
	III	210	188.25	1.1155	1.135	0.983
	IV	185	191.375	0.9667	0.916	1.055
1981	I	178	194.0	0.9175	0.936	0.980
	II	201	195.125	1.0301	1.019	0.908
	III	214	197.625	1.0829	1.135	0.954
	IV	190	200.75	0.9465	0.916	1.033
1982	I	193	203.375	0.9490	0.936	1.014
	II	211	206.0	1.0243	1.019	1.005
	III	225			
	IV	200			

Once again, these should not be "going anywhere" and in this case, that means that the arithmetic mean of them should be 1. In fact, it is 1.0019 to 4 d.p. and so, for accuracy, we shall adjust the components to be (0.936, 1.019, 1.135, 0.916).

To interpret these, we must remember that we are employing a multiplicative model. Thus, the first quarter's figures are, on average, 6.4% below the trend, the second quarter's figures are, on average, 1.9% above the trend, and so on.

(ii) By geometric mean, we multiply the relevant numbers together and take the nth root. In this case, we multiply the four numbers together and take the fourth root. (If you have a calculator which does not allow you to take the fourth root directly, you can find out the fourth root by depressing the square root button twice.)

For quarter I, the estimated component is

$$(0.4387 \times 0.9392 \times 0.9175 \times 0.9490)^{1/4} = 0.9360$$

For quarter II, the estimated component is

$$(1.0274 \times 0.9973 \times 1.0301 \times 1.0243)^{1/4} = 1.0197$$

For quarter III, the estimated component is

$$(1.1936 \times 1.1501 \times 1.1155 \times 1.0829)^{1/4} = 1.1348$$

For quarter IV, the estimated component is

$$(0.8430 \times 0.9081 \times 0.9667 \times 0.9465)^{1/4} = 0.9148$$

To check that these are stationary, we multiply them together and get 0.9908 instead of 1. We can, therefore, adjust them to (0.938, 1.022, 1.138, 0.917).

You can see that there is very little difference between the two methods as regards the actual results obtained. However, three points ought to be made here. Firstly, the arithmetic mean will always result in a larger number for the estimated component than the geometric mean. However, adjustment to give a mean of 1 will obscure this. Secondly, the geometric mean approach is probably far better from the purist technical point of view. However, as the differences are small, the easier method will, in general, be adopted. Thirdly, large differences arise if the actual numbers from the detrended series vary wildly. For example, if the detrended figures for quarter 1, had been 0.71, 0.94, 1.13, and 0.75, then the arithmetic and geometric means are 0.8825 and 0.8672, respectively. If such a difference occurs and is felt to be significant, attention ought to be paid to the whole model and to the actual data which gave rise to such differing numbers rather than the method of estimating the seasonal components.

We can now enter the (arithmetic mean) seasonal components in Table 14.4 and, by division, produce a detrended, deseasonalized series.

14.5.3   *Multiplicative Model—The Cyclical Components*

For the same reasons as when we discussed the additive model, we shall not estimate the cyclical component. However, a little thought would lead us to the conclusion that the cyclical component may itself have to increase (or decrease) in magnitude if the series is a long one. If the economy is growing then the cycles may cause a rise (fall) of 3%, say. A percentage increase is far more appropriate than an increase in absolute terms.

14.5.4   *Multiplicative Model—The Residual Component*

As with the additive model in §14.3.4, we must have some idea of the residue in this model. Once again, a simple idea is to evaluate the mean and variance and look for odd values.

With this model, the mean is 0.9938125 and the variance is 0.001658 and, therefore, the standard deviation is 0.0407. Now, compared with the residual component in the additive model, these figures are very small. However, these are measured in different units. The residuals in the additive model are measured in units of house sales. These residuals are measured as factors. Thus, the standard deviation is 4.07%.

14.5.5 *Multiplicative Model—Forecasting*

A rough line can be fitted by the method of semi-averages. In fact, it will be the same line as in 14.3.3. This has equation

$$\text{Trend} = 156.085 + 2.826 \text{ (Number of quarter)}$$

We can now forecast the 1983 figures.

1st quarter 1983
   Estimated trend value is 215.431
   First quarter seasonal component is 0.936
Thus, our estimated figure for the first quarter of 1983 is 201.640.
   For the second quarter of 1983, the estimate is 218.257 × 1.109 which equals 222.40.
   For the third quarter of 1983, the estimate is 221.083 × 1.135 which equals 250.93.
   For the fourth quarter of 1983, the estimate is 223.909 × 0.916 which equals 205.10.

14.6 Additive Model v Multiplicative Model

Although many other models are possible, we have dealt with only two. However, we have not really discussed when to use each model. There is no simple answer as neither of them are, in any sense, correct. It is merely a question of which one provides an adequate and appropriate model. Two things ought to be borne in mind here. The first is that a model is of little use if its results cannot be interpreted in the real world. This may affect our decision on which model is more appropriate. The second thing is that most economies tend to react in percentage, i.e. relative, terms and not in absolute terms. Thus, particularly if we have a long time series, an additive model may be difficult to justify.

14.7 Fitting a Trend Line by a Resistant Line

In chapter 11, we mentioned the idea of robustness. To recap, a robust statistical procedure is one which is not too sensitive to our assumptions. Such a procedure is the method of fitting a resistant line.

Roughly speaking, a resistant line is found by ordering the data by use of the co-ordinates. Then, we split the data into three groups of about the same size. We find a point to represent each of the three groups and draw a line between the outer two points. We use the proximity of the line and the middle point to verify the adequacy of a straight line.

If we refer to the centred moving average of Table 14.2 above, we can order the data in time. We have sixteen points and so we can split the data up into three groups as follows.

Group 1		Group 2		Group 3	
Quarter	C.M–A.	Quarter	C.M–A.	Quarter	C.M–A.
3	163.375	8	179.5	14	195.125
4	167.25	9	181	15	197.625
5	169.375	10	184.5	16	200.75
6	173.25	11	188.25	17	203.375
7	177.375	12	191.375	18	206
		13	194		

For a representative point for each group, I shall use the median. (Recall that the median is more robust than the mean.) Thus, the three points are $(5, 169.375)$, $(10.5, 186.375)$, and $(16, 200.75)$. Note that the outer two representative points are actual points in the centred moving average. This is due to the smoothing which often gives data in strict order. This need not always be the case.

The line through the two outer points has equation

$$\text{Trend} = 155.1137 + 2.85227 \text{ quarter}$$

Now, if the quarter is 10.5, the figure on the trend ought to be 185.0625. This is quite close to the 186.375 we actually have. If it is far away, then it would have suggested that there is some curvature and, therefore, a straight line is not an adequate model.

Resistant lines are increasing in use because of their robustness and their simplicity relative to the classical regression line with which we deal in the next chapter. Firstly, however, we ought to discuss what to do if the trend is curved.

14.8 Fitting Curves

In many situations in statistics we want to represent data which seem to have a curved shape, with a smooth curve for the purposes of policy-making, forecasting, interpretation, etc. Fitting curves is not, however, a simple matter as, firstly, we must decide on the form of the curve and, secondly, we must decide on the method to be used to solve for the parameters.

For example, suppose we have data which are getting larger year by year. If we let T denote the line and X denote the data we could then consider adopting the following models. Other Capital letters denote parameters.

(i) $X = Ae^{BT}$
(ii) $X = A + BT + CT^2$
(iii) $X = A + BT + CT^2 + DT^3$
(iv) $X = A \ln T + B$
(v) $X = AT^2 / BT + C$
(iv) $X = Ae^T T^B$

Many others are possible and, perhaps, it is the ability of the trained statistician which allows him/her to, at least, rule out many inappropriate models.

As regards the method of solution for the parameters, many mathematical methods exist which shall not be discussed here. However, one suggestion which is of use in some practical situations is to take logarithms.

In the above models, by taking logarithms,

(i) becomes $\ln X = \ln A + BT$ and
(vi) becomes $\ln X = \ln A + T + B \ln T$.

Particularly in the former of these, we have reduced the model to a very simple form for if we take logs of all the data and call this L and rewrite $\ln A$ by G we have a simple equation of $L = G + BT$, i.e. we have returned to fitting a straight line but to the logarithms of the data.

Similarly, an alternative way to deal with a multiplicative model is to take logs. If we believe that $A = T \times S \times C \times R$ is an appropriate model, then so must be $\ln A = \ln T + \ln S + \ln C + \ln R$, i.e. an additive model but in the logarithms of the data and of the components. We must remember to take exponentials before trying to interpret the components.

Most calculators and sets of mathematical tables will take logs for you. The only problem occurs if any of the data is either negative or positive but very small. In this case, it may be possible to make slight adjustments to the data and/or the model so that we can analyse the data.

14.9 Exponentially Weighted Moving Averages

When we discussed simple moving averages in §14.2.2, the moving average for a particular point in time is calculated in such a way that this point in line and adjacent points in time carry the same amount of weight.

An alternative approach is to use the current point in time to have the largest weight, the previous point in time the second largest weight and so on. In effect, the weight falls off or decays. This decay can be thought of as a correlation. For example, suppose the decay factor is 0.4. If we take the data in Table 14.2, we can construct an exponentially weighted moving average.

Simple algebra shows that, if we denote the moving average for time t by MA_t and the datum for time t by A_t, then, as we have a decay factor of 0.4,

$$MA_t = 0.4 A_t + (1 - 0.4)MA_{t-1}$$

The size of the decay factor has to be decided in each case. This factor can be fixed at any value between 0 and 1 but the lower it is, the greater the smoothing. While we are trying to smooth the data, we must take care not to smooth out all the effects and components in which we are interested. If we have a computer or programmable calculator available, the answer is usually to try several different values of the decay factor to see which is most appropriate. Alternatively, we may have a prior belief about the appropriate value of this decay factor.

Another problem is how to obtain the first figure without a previous figure to enter in the formula above. The simplest way is to calculate a simple mean of, in this case, the first year or two of data. As a rough figure, however, we can use 158.9 which is roughly the value we get from the trend line calculation in §14.4.3. The exponentially weighted moving average appears in Table 14.5 below under the column marked 0.4.

The fact that the figure for the first quarter of 1978 is approximate is of little importance for, if we had a figure of 0 for this point in time, the moving average value for the first quarter of 1980 is still 172.7, i.e. the effect of a drastic change in one of the figures is smoothed out within 8 figures.

We should note that this moving average is not as smooth as the one calculated in §14.2.2. Perhaps, a decay factor of 0.2 would be better. The corresponding exponentially weighted moving average appears in Table 14.5. This is a lot smoother but still retains the basic seasonal fluctuations. Perhaps a decay factor of 0.15 or 0.1 should be tried.

However, the important point to note is that there is more than one way to smooth the data. Choice of method depends upon the context and the data, and the calculating power available.

Table 14.5

		Actual data	Exponentially Weighted Moving Average	0.4	0.2
1978	I	136		158.9	158.9
	II	170	$0.6 \times 158.9 + 0.4 \times 170$	163.3	161.1
	III	195	$0.6 \times 163.3 + 0.4 \times 195$	176.0	167.9
	IV	141	$0.6 \times 176.0 + 0.4 \times 141$	162.0	162.5
1979	I	159		160.8	161.8
	II	178		167.7	165.1
	III	204		182.2	172.8
	IV	163		174.5	170.9
1980	I	170		172.7	170.0
	II	184		177.2	173.4
	III	210		190.3	180.7
	IV	185		188.2	181.5
1981	I	178		184.1	180.8
	II	201		190.9	184.9
	III	214		200.1	190.7
	IV	190		196.1	190.6
1982	I	193		194.8	191.0
	II	211		201.3	195.0
	III	225		210.8	201.0
	IV	200		206.5	200.8

Examples
1. The production of passenger cars in a country is recorded for 4 years as follows:

Year	Quarter	Production (in thousands)
1972	2	159.6
	3	168.8
	4	173.2
1973	1	141.8
	2	138.7
	3	159.2
	4	152.0
1974	1	118.2
	2	136.3
	3	139.5
	4	132.1
1975	1	124.4
	2	125.9
	3	137.4
	4	120.3
1976	1	106.0

Analyse this time series using an appropriate model and estimate production levels in

(a) third quarter of 1976
(b) calendar year 1977
(c) first quarter of 1980

How accurate are your estimates of production and what assumptions are being made?

2. The turnover (in £000's) for a Glasgow restaurant is recorded. For this set of data, use an appropriate model to estimate turnover in calendar year 1984 and in the autumn of 1986. State clearly any assumptions you are making.

	Spring	Summer	Autumn	Winter
1980		185	183	204
1981	216	261	272	311
1982	351	394	375	425
1983	440	482		

3. Data on daily city bus traffic (in thousands of passengers) is recorded for the weekdays for a 3-week period as below.

	Week 1	Week 2	Week 3
Mon	182	168	153
Tue	153	139	122
Wed	171	156	143
Thu	138	117	112
Fri	174	158	147

Use an appropriate model to estimate the daily effects.

4. Office maintenance costs (in £'s) are recorded as follows

	1978	1979	1980	1981
Jan–Feb	721	805	918	1023
Mar–April	758	827	927	1047
May–June	626	681	743	
Jul–Aug	704	764	833	
Sep–Oct	731	802	885	
Nov–Dec	790	857	929	

Estimate costs for Sept–Oct 1981, for March–June 1982, and for calendar year 1984.

5. Quarterly imports of merchandise by a western country are recorded (in millions of national currency units).

	1977	1978	1979	1980	1981
Jan–Apr	109	138	160	186	211
May–Aug	120	147	171	195	219
Sep–Dec	153	182	214	243	285

Using an appropriate model, estimate the imports for Jan–April 1983 and for Sept–Dec 1984. State clearly any assumptions you are making.

6. As part of its service to the community, a University runs a 10-lecture series on Computing. These lectures are in the evenings, are free, and are open to the public. The series runs in each of the University's three terms and the numbers of people who enrolled are recorded for five years.

	1978–79	1979–80	1980–81	1981–82	1982–83
Term 1	7	13	18	29	44
Term 2	18	26	35	47	72
Term 3	8	11	16	20	33

Use an appropriate model to forecast the enrolment figures for each term of the 1983–84 academic year.

CHAPTER 15

Ordinary Least Squares Regression

15.1 Introduction

In the previous chapter, we discussed some methods for fitting a line to a set of points. The reason there are a number of methods to choose from is that, as we try to decide on a line that is closest to the points, we cannot usually decide how to measure closeness.

Take, for example, the problem of going from London to Edinburgh. There are various methods of transport. Which one is best? We could walk, cycle, or go by car, travel by bus, taxi, train or aeroplane. With each of these methods, we can evaluate a probability distribution for the time taken for the journey. This will reflect not only the speed of the mode of transport but also its reliability. In addition, we need to consider costs and comfort as well as other factors. That this problem does not have a simple solution is evident by the fact that several of these modes of transport are available commercially.

A similar problem exists in deciding on a best line. Several mathematical approaches are possible. In developing the Ordinary Least Squares Regression line, we shall mention other possible lines in addition to those already dealt with in the previous chapter.

15.2 Ordinary Least Squares Regression Line—Mathematical Development

As an example to discuss, we shall consider the following set of data. They consist of the import data for various ports of the UK. In fact, the data are recorded for twenty-eight ports and are the total tonnage of imports (recorded in thousands of tonnes) and value of imports (recorded in millions of pounds sterling).

As can be seen from the data, the points do fall close to a straight line. Clearly a straight line would be useful as it would allow us to make reasonable predictions of the value of imports from the tonnage as well as allowing us to examine the relationship between these two variables.

To decide on a specific line we must decide on what we mean

Table 15.1

Port	Tonnage (000's of pounds)	Value (£m's)
1	11,304	988
2	295	396
3	1,003	203
4	4,731	398
5	10,073	1,907
6	1,722	107
7	1,150	282
8	3,146	448
9	13,136	922
10	8,550	1,120
11	612	266
12	933	597
13	13,104	2,113
14	2,256	273
15	413	127
16	6,314	721
17	9,544	982
18	10,348	1,493
19	11,025	1,731
20	8,875	803
21	8,038	1,022
22	4,202	804
23	8,605	1,255
24	11,204	1,486
25	4,188	602
26	7,813	631
27	626	1,026
28	4,302	922

by a line being closest to the points. If we draw a line through this graph, then we must first measure how far a point is from the line. One way to measure this distance is the vertical distance. Another is the horizontal distance while a further possibility is the perpendicular distance. All three are possible but we shall use the vertical distance. This is done for mathematical reasons which, in fact, result in the line we eventually derive being what is called scale invariant. In other words, taking the vertical distance means that it does not matter if we record the value of imports in millions of pounds or thousands of pounds or billions of yen. The other distances would not lead to scale invariance as the actual line we derive will be dependent on the scale of the variables.

Another decision that has to be made is what we mean by closest. In other words, how do we get a small distance between the points

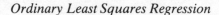

Fig 15.1

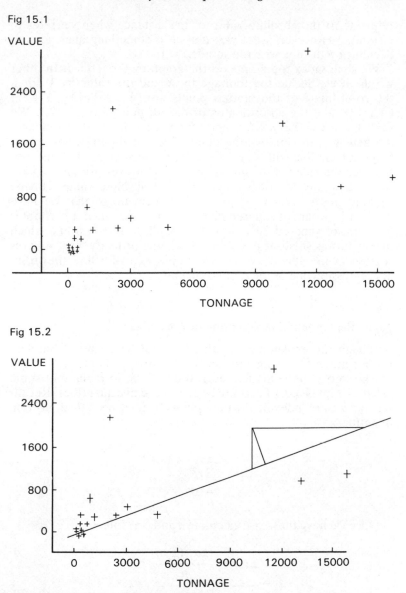

Fig 15.2

and the line? One suggestion is to decide on the line which has a maximum distance as small as possible. Other possibilities are to make the mean distance as small as possible or the median distance as small as possible. (Note that in most cases, we are referring to the absolute value of the distance when we refer to distance.) However, what we shall do is something analogous to deciding on the line with the smallest variance.

We shall index the points on the graph by the letter i. In other words, if we decide the tonnage by X and the value by Y, then the co-ordinate of the plotted points are (X_1, Y_1), (X_2, Y_2) ... (X_n, Y_n), and the co-ordinates of the ith point are (X_i, Y_i). We shall fix a line $Y = a + bX$ and, for each value of the tonnage, the height of the line will be denoted $\hat{Y}$; for the ith tonnage, the height of the line will be $\hat{y}_i$. For each tonnage value, the distance between the value of the line is $|y_i - \hat{y}_i|$. However, we shall square this and so we do not need to take the absolute value. Having squared this to get $(y_i - \hat{y}_i)^2$, we add them up so that we have $\Sigma(y_i - \hat{y}_i)^2$, a sum of squared distances or as we called it in chapter 5, a sum of squared deviations. We shall decide on a line which makes this as small as possible. This is the ordinary least squares regression line. It is also called the simple regression line, the ordinary regression line or the regression line.

15.3 Regression Line—Arithmetic Formulae

Although the explanation of this line may seem quite complex, the formulae for fixing the line are relatively simple. (We shall not go into a mathematical derivation of the formulae but some calculus is involved.) To fix the line, we need five quantities. Recalling that we are indexing the data points by i, we need the following quantities:

$$\Sigma x_i$$
$$\Sigma y_i$$
$$\Sigma x_i^2$$
$$\Sigma x_i y_i$$
$$n$$

Once we have these, we calculate a and b as follows

$$b = \frac{\Sigma x_i y_i - \dfrac{(\Sigma x_i)(\Sigma y_i)}{n}}{\Sigma x_i^2 - \dfrac{(\Sigma x_i)^2}{n}}$$

$$a = \frac{\Sigma y_i}{n} - b\frac{\Sigma x_i}{n}$$

For the data we are considering

$$n = 28$$
$$\Sigma x_i = 167{,}512$$
$$\Sigma x_i^2 = 1{,}508{,}457{,}472$$
$$\Sigma y_i = 23{,}625$$
$$\Sigma x_i y_i = 191{,}949{,}760$$

Thus, $b = \dfrac{191{,}949{,}760 - (23{,}625)(167{,}512)/28}{1{,}508{,}457{,}472 - \dfrac{(167{,}512)^2}{28}}$

$$= 0.09996$$

and

$$a = \frac{23{,}625}{28} - 0.09996 \times \frac{167{,}512}{28}$$

$$= 245.732$$

For simplicity's sake, we shall take this line to be $Y = 245.7 + 0.1X$.

Thus, we are saying that the line which minimizes the sum of squared deviations has equation $Y = 245.7 + 0.1X$. In addition, we need to have three other quantities. These are the sums of squares. The first one we have already discussed and this is the sum of the squared deviations or, more correct statistically, the sum of squared errors or sum of squares due to error. This is often denoted SSE, and is equal to $\Sigma(y_i - \hat{y}_i)^2$, the very quantity we are trying to minimize. Another important sum of squares is the total sum of squares, usually denoted TSS. This is equal to $\Sigma(y_i - \bar{y})^2$, the sum of squared deviations of the value figures from their mean. The third sum of squares is the sum of squares due to the regression or the regression sum of squares. This is usually denoted RSS and is equal to $\Sigma(\hat{y}_i - \bar{y})^2$. However, only two of these need to be evaluated as

$$TSS = RSS + SSE$$

In our case, $TSS = 7{,}906{,}211$, $RSS = 5{,}059{,}251$ and $SSE = 2{,}846{,}960$.

15.4 Regression Terminology

In classical economic texts, the X variable is denoted the dependent variable and the Y variable is denoted the independent variable.

However, such terminology is highly misleading as the variables are not independent of each other. Instead, it is much clearer to refer to the X variable as the explanatory variable or carrier variable and the Y variable as the response variable. Another expression is the term "the regression of Y on X" or, in the case, "the regression of value on weight". This is merely a brief way of stating which is the explanatory and which the response variable.

The height of the line at each X value is denoted Y and will be referred to as the predicted Y-value. Further, we shall look at the difference between the predicted values and the actual values. As in the previous chapter, these are called the residuals and will be denoted by $(y_i - \hat{y}_i)$ or by e_i, where e is used as these are also called error terms. (Note that in economic theory, these residuals are often called disturbance terms.)

15.5 Regression—The Statistical Model

We have glossed over the statistics which makes this line the "best" line. We must now look at the underlying model and its assumptions.

15.5.1 The Model and its Assumptions

The model is that

$$Y = \alpha + \beta X + e$$

where as before we denote the population parameters by Greek letters. In estimating α and β by a and b, we make the following five assumptions.

(1) The values of the explanatory variable are known exactly.
(2) $E(e) = 0$
(3) The residuals have an approximate Normal distribution.
(4) The variance of the residuals is approximately constant.
(5) The residuals are independent of each other.

We can now look in turn at these assumptions. The first one is crucial and is often omitted in textbooks. The model requires that the explanatory variable is known exactly. It cannot be subject to a significant measurement or recording error. If it is, then we no longer are involved in regression methods or regression analysis but have to move into an area of econometrics called specification analysis or simultaneous equation regression analysis. There is no way to mathematically check that this assumption holds. Rather, it is a matter of understanding the context and the source of the data.

The second assumption requires that the residuals are distributed about zero. If they are not, then this would probably indicate that some factor has been omitted. We shall briefly discuss how to use two or more explanatory variables in the next chapter. However, if we fit a line of the form $Y = \alpha + \beta X$, then we are mathematically guaranteed to have a zero mean for the residuals. Our problems may occur if we specify, for example, that the line must go through the origin, i.e. have a zero intercept. In other words, if we specify that α must be zero and fit $Y = \beta X$, we are no longer guaranteed a zero mean for the residuals.

The third assumption is required later when we perform some hypothesis tests on the results of the regression. If the residuals appear to follow some other distribution, then much of the analysis is invalidated. Several tests for Normality exist. A cursory test is to construct a stem-and-leaf display or an ogive of the residuals and examine if they resemble the theoretical shape we would get from an exact Normal curve.

We also make the assumption that the variance is constant. This is not always the case. In some data sets, the variance increases as the size of the explanatory variable increases. Thus, if we plot the residuals against the explanatory variable, we may get a share similar to that in Figure 15.3. This is a case of heteroscedasticity, i.e. unequal variances. What we want is homoscedasticity, i.e. equal variances. Once again, if we have unequal variances but

Fig 15.3

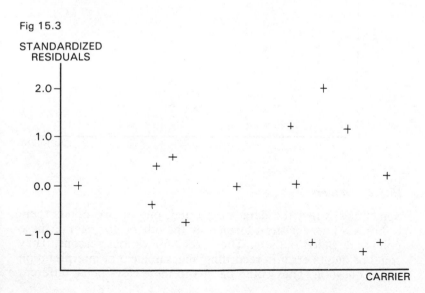

STANDARDIZED
RESIDUALS

CARRIER

know something about the structure of this inequality, we may be able to introduce this knowledge into our analysis.

The fifth assumption is quite a tricky one. It is very difficult to prove independence and, therefore, we usually examine plots of the residuals against the explanatory variable, residuals and the response variable, residuals against each other, i.e. the ith residual against the i-1th residual, and residual against i. Any pattern in these plots can be used as evidence either of dependence or of inappropriateness of the model.

Patterns such as these in Figure 15.4a and Figure 15.4b are evidence of something odd. The first one suggests that some substantial change occurred in the middle. We ought to investigate the cause and nature of this change and, if possible, estimate it and include it in our analysis. The second figure suggests that the original data should not be modelled by a straight line. It appears that some type of curve is more appropriate.

Fig 15.4a

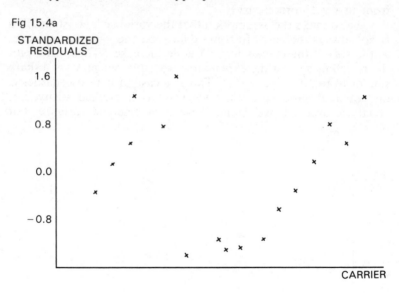

15.5.2   *Outliers*

Sometimes, when the data are plotted, one or two points seem to break a linear pattern formed by the others. In general, these points are called outliers. They occur for various reasons. They could be due to error in recording, measurement or interpretation of the question. They could be due to these cases being different

Fig 15.4b

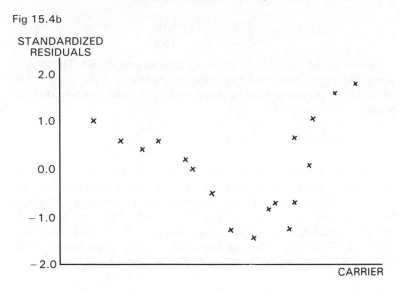

in nature from the other points. If the former is the cause, a correction should be attempted. If we acknowledge that these cases are different, we can either introduce this difference into the model, or omit these points from consideration. This latter solution is often the easier but is not always the better. In general, something has to be done as these outliers may spoil or obscure whatever structure there is but great care must be exercised.

15.6   Regression—Variation Explained

We can measure how much benefit we get from the regression line. Suppose that we are to use the regression line to predict value given information on tonnage. If we do not use the information on tonnage, the best estimate of value for a port is the mean of the values of the ports which we already have. If we do this, the errors involved can be measured as TSS or $\Sigma(y_i - \bar{y})^2$. Now, if we use the information on tonnage, the errors involved in these predictions are measured as SSE or $\Sigma(y_i - \hat{y}_i)^2$. Thus, we have an expression for the errors we incur if we do the best we can but without information on tonnage as well as an expression for the errors we incur if we do the best we can but have the information on tonnage. If we divide one by the other we get an expression for the proportion of variation which remains, i.e. cannot be accounted for or removed by the information on tonnage.

$$\frac{\Sigma(y_i - \hat{y}_i)^2}{\Sigma(y_i - \bar{y})^2} = \frac{SSE}{TSS}$$

However, we generally calculate a quantity called R^2. This represents the "proportion of variation explained" or the proportion of variation accounted for or removed by the information on tonnage.

$$R^2 = 1 - \frac{\Sigma(y_i - \hat{y}_i)^2}{\Sigma(y_i - \bar{y})^2} = 1 - \frac{SSE}{TSS} = \frac{RSS}{TSS}$$

Now, if we have a simple regression with only two variables, R^2 is also equal to the square of the correlation between the two variables. In this case, the only other piece of information which would be needed is Σy_i^2 which is 27,839,805. Thus, we can work out the correlation to be 0.8. Accordingly, R^2 is 0.64. In other words, 64% of the variation incurred if there is no information is accounted for if we we have information on tonnage.

15.7 Estimate of the Error Variance

We need to have some idea of how much error there is remaining. In other words, what is the variance of the residual component. We estimate this by $SSE/n - 2$. (We use $n - 2$ instead of $n - 1$ because, in technical terms, we use two degrees of freedom to estimate the two parameters of the model.)

15.8 Regression—Tests of the Model

One of the main uses of regression is to analyse the structure of a relationship. Thus, in this example, we know that there is some relationship between tonnage and value but we wish to know what it is. The most fundamental step is to carry out one or more hypothesis tests.

15.8.1 Hypothesis Test of the Significance of β

The most common test is of the significance of β. In other words, we are testing whether β is significantly different from zero, i.e. if the line has any significant slope. We can set up a formal hypothesis test.

$$H_0: \ \beta = 0$$
$$H_1: \ \beta \neq 0$$

5% significance level

Under H_0, b is approximately Normally distributed with a mean of $\beta(= 0)$ and a variance of

$$\frac{SSE/n - 2}{\Sigma(x_i^2) - \dfrac{(\underline{\Sigma x_i})^2}{n}}$$

Thus, our test statistic has a value of $0.1/0.0002163$ which equals 6.80. Thus, we reject H_0 and conclude that the shape of the line is significantly different from zero. Even although our estimate of the shape is 0.1, we can conclude that the true value of the shape is significantly different from zero.

15.8.2 *Hypothesis Test of a Specified Value for* β

We are also in a position to test whether a specified value of β is being maintained. Suppose, for example, previous evidence had shown that the slope of the line ought to be 0.04 but recent changes in foreign trade administration had possibly altered the structure. We can test whether this value of 0.04 is being maintained.

$$H_0: \ \beta = 0.04$$
$$H_1: \ \beta \neq 0.04$$
5% significance level

Under H_0, $b \sim N(0.04, 0.0002163)$

H_0 is rejected if the test statistic b falls outside the region $0.04 \pm 1.96 \times 0.01471 = (0.0112, 0.0688)$. As the estimate b is 0.01 which does not fall in this region, we must reject H_0 and conclude that the previous slope of 0.04 has not been maintained.

15.8.3 *Hypothesis Test of* α

Sometimes, we are interested in making inferences about the value of α, the intercept of the line.

We can set this up as a formal hypothesis test.

$$H_0: \ \alpha = \alpha_0$$
$$H_1: \ \alpha \neq \alpha_0$$
5% significance level

We are looking at the general case. In most cases, α, would be zero.

Under H_0, our estimate a is approximately Normally distributed with mean α_0 and variance equal to

$$\sigma^2 \left(\frac{1}{n} + (\Sigma x_i)^2\right) \Bigg/ \left[n^2\left(\Sigma(x_i^2) - \frac{(\Sigma x_i)^2}{n}\right)\right]$$

where, once again, σ^2 is estimated by $SSE/n - 2$. Thus, if we are to test whether the intercept is significantly different from zero, we set $\alpha_0 = 0$.

Under H_0, $a \doteq N(0, 11{,}651.22)$.

We reject H_0 if a falls outside the range $0 \pm 1.96 \times \sqrt{11{,}651.22} = (-211.56, 211.56)$. As a equals 245.7, we can conclude that the y-intercept of the line is significantly different from zero.

15.8.4 A note on Hypothesis Tests

In carrying out more than hypothesis test on a set of data, we have some degree of dependence and also, because of the number of tests, we may get a spurious result. We must be careful here and the reader is referred to §16.2.2 for further explanation of the problem.

15.9 Predicted Values

One of the two main reasons for embarking upon a regression analysis is to be able to make predictions about values of a variable. However, before we proceed to that, we must make a very important distinction. In our example above, we have assumed a model of $Y = \alpha + \beta X + e$ where $e \doteq N(0, \sigma^2)$. What this means is that, at some value of tonnage X_0, the corresponding monetary value Y has a Normal distribution with a mean of $\alpha + \beta X_0$ and a variance of σ^2. We may have several ports with a tonnage of X_0, say. We must be clear as to whether we are trying to predict the mean monetary value of the imports at these ports or the monetary value of the imports at one particular port. In other words, we must distinguish between the prediction of a mean response for a given value of the explanatory variable and the prediction of an individual response.

15.9.1 Prediction of a Mean Response

We are going to construct a confidence interval for a mean response. The location of this confidence interval will depend upon the parameter estimates but the width depends upon the amount of information we have and its accuracy. The more information we have and the more accurate it is, the narrower will be the confidence interval. Thus, if we have a very small residual variance, the confidence interval will be narrow. If we are trying to predict a mean response for a value near which we already have a lot of information, the confidence interval will be narrow. The converse is true so that if there is a large residual variance or if the prediction is for a value far from our other values, the interval will be wide.

We shall consider a prediction of monetary value for a tonnage of Xc. The confidence interval will be centred on $\alpha + \beta Xc$ which we estimate by $a + bXc$ and the variance of this estimate is

$$\sigma^2\left(\frac{1}{n} + \frac{(Xc - \bar{x})^2}{\Sigma x_i^2 - \dfrac{(\Sigma x_i)^2}{n}}\right)$$

Again, σ^2 is estimated by $SSE/n - 2$.

Thus, if we are to estimate the mean monetary value for ports with imports of 700,000 tonnes, we can construct a 90% confidence interval as

$$245.7 + 0.1 \times 700 \pm 1.645 \times \sqrt{\frac{2{,}846{,}960}{26}\left(\frac{1}{28} + \frac{(700 - 843.75)^2}{\Sigma x_i^2 - (\Sigma x_i)^2/n}\right)}$$

$$= 245.7 + 70 \pm 1.645 \times \sqrt{(109{,}498.46 \times 0.035755)}$$
$$= 315.7 \pm 1.645 \times \sqrt{3{,}915.1174}$$
$$= 315.7 \pm 102.929$$
$$= (212.8, 418.6)$$

You will notice that the width of the interval is affected by two additional factors. If the tonnage values for which we already have information are widely spread, then so will be $\Sigma x_i^2 - \dfrac{(\Sigma x_i)^2}{n}$, i.e. if we have a wide spread of information, then our confidence interval will be narrower and our prediction will be more accurate. Also, if the sample size n is increased so that we have more information, another increase in accuracy of prediction will occur.

15.9.2 *Prediction of an Individual Response*

This approach is very similar except that the interval must be somewhat wider. As a result, the variance of the estimate is now

$$\sigma^2\left(1 + \frac{1}{n} + \frac{(X_i - \bar{x})^2}{\Sigma x_i^2 - (\Sigma x_i)^2/n}\right)$$

i.e. we have added a "1" in to the expression. If we are trying to predict the monetary value for an individual port with a tonnage of 700, our 90% interval is now,

$$315.7 \pm 1.645\sqrt{109.498.46 \times 1.035755}$$
$$= (-238.3, 869.7)$$

As you can see, this interval is much wider and is of relatively little use. However, that is not to say that this confidence interval is useless. The message here is that there is not enough relevant accurate information on which to base a prediction of an individual response.

15.9.3 Problems with Prediction

As in the time series analysis in the previous chapter, we must be very careful about making predictions far beyond our range of experience. For example, suppose we had a port which handled only ten tonnes of imports last year. The regression model would produce a point estimate of a value of $245.7 + 0.1 \times 0.01 = £245.701m$. However, common sense should suggest that, for such a low tonnage, the goods must be very valuable, otherwise the port could never make a profit. For example, this port may specialize in imports of diamonds whose value would be well in excess of £245m. Accordingly, caution and common sense must be exercised in applying the structure of the regression model beyond our range of experience and information.

Another problem occurs with causality. At no point in this chapter have we stated that a particular value of the X-variable causes a particular value of the Y-variable. We are merely saying that there is an association statistically, that the values tend to occur together to form a pattern. *It is not possible for statistics to prove that one thing causes another.*

15.10 Regression on Computers

Most modern software packages for business analysis or statistical analysis will contain facilities for performing regression. In this way, the tedious calculations are removed. However, one must still understand what is going on so as to be able to interpret the results. Perhaps, the best widely available package for non-specialists which has a regression facility is MINITAB.

Examples
1. The following data show, for 1980–1981, the median weekly household income and the median weekly household expenditure on commodities and services for 15 families in one region of the UK.

Median weekly household income (£)	Median weekly Expenditure on Commodities etc. (£)
140.24	111.22
135.01	103.98
148.63	114.78
151.65	113.79
183.59	135.18
149.06	113.86
155.37	119.15
155.43	116.15
143.81	110.54
140.81	112.86
176.35	127.42
122.53	102.31
161.55	128.44
183.40	131.58
170.36	130.66

Fit a regression line of expenditure on income. At the 5% level, test the significance of the slope. Predict the mean median expenditure for families with a weekly income of £147.09. Give a 95% confidence interval.

2. The following data are recorded for 10 industrial countries. For each country, there is the 1982 volume of exports, the national interest rate, and the population.

Country	Exports ($m)	National Interest Rates	Population (m)
Portugal	4,165	19.00	10.06
Denmark	15,309	10.00	5.12
Sweden	26,786	10.00	8.33
Iceland	685	28.00	0.227
Jamaica	710	11.00	2.22
Switzerland	26,019	4.50	6.47
Canada	71,239	10.05	24.34
Netherlands	66,314	5.00	14.25
South Africa	17,694	13.50	30.13
Japan	138,403	5.50	117.65

Fit a regression line so as to be able to predict exports from interest rate. Fit another regression line so as to be able to predict exports from population.

Clearly, exports and population are closely linked as a country's population capacity is related to its population. A better variable to predict might be per capita exports. Fit a regression line so as to be able to predict per capita exports from interest rates.

3. A large manufacturer has hired an economist to investigate the relationship between family income and the purchase price of new cars. Accordingly, a sample of 14 people purchasing new cars was selected and the data recorded were as follows

Family Income (£000's)	Car Price £000's)
10.3	4.2
16.2	5.3
21.6	6.4
18.8	6.2
19.4	6.5
14.5	5.2
27.3	7.9
18.0	5.7
23.2	7.4
17.1	5.8
16.9	5.5
35.6	10.3
22.7	6.6
15.6	5.1

Fit a regression line so as to be able to predict car price from family income.

Test the significance of the intercept.

Predict the price of a car bought by someone with a family income of £14,900. Give a 95% confidence interval.

Now, fit a regression line of family income on car price.

These two lines have different sets of assumptions. Which set do you think is more likely to be a valid set?

CHAPTER 16

Multiple Regression

16.1 Development from Simple Regression

The idea of multiple regression is that we can examine more than two variables at a time. For example, in the case we discussed in the previous chapter, we may have information on the number of ships handled in a year, annual charges for tug-boats, annual cost of port authority vehicle maintenance, etc.

Instead of fitting the "best" line to a two-dimensional set of points, if we have three variables, we shall be fitting the "best" plane. In other words, we have a three-dimensional graph and move around a flat piece of paper until it is in the "best" position. If we have four variables, then we are, unfortunately, into higher dimensions than we can physically see. The principle, however, is the same.

The mathematical formulae are not simple and will not be discussed here. There is a large range of software which can perform the calculations simply (and painlessly) for us.

Let us suppose that we have five variables. Four of them are explanatory variables and one is the response variable. The model we would be fitting is

$$Y = \alpha + \beta_1 X_1 + \beta_2 X_2 + \beta_3 X_3 + \beta_4 X_4 + e$$

and the estimates a, b_1, b_2, b_3, and b_4 will be chosen so as to minimize $\Sigma(y_i - y_i)^2$. This model has the same assumptions as those in 15.5.1.

Now, if R^2 measures the proportion of original variation which is explained or removed by the model, it should be clear that every time we introduce another variable into the model, R^2 cannot decrease and will, in almost all cases, increase. If the new variable is of no help in reducing the variation, the parameter estimate will be zero. Thus, we must be careful. If we want to have R^2 close to 100%, we can achieve this by introducing as many variables as we can. These need not be sensible variables but sooner or later R^2 will approach 100%.

As a result, we must deal with a slightly different measure and produce an adjusted R^2. This can, in fact, decrease if we introduce a variable which is of minimal use.

As with simple regression we can perform hypothesis tests of and/or construct confidence intervals for the parameters. These are of similar form to those in the previous chapter although it is possible, for example, to perform a test involving three parameters. For example, we could specify H_0: $\beta_1 = \beta_2 = \beta_3 = 0.4$ or H_0: $\beta_4 > \beta_1 > \beta_3$.

16.2 Multiple Regression—New Issues

There are some issues which arise only when we move from simple to multiple regression. The interpretation and use of R^2 has been mentioned above but three other serious issues need to be discussed here.

16.2.1 Multicollinearity

Ideally, we would like the explanatory variables to have a zero correlation between each other. This is very rare. Usually we settle for low correlations. If we do not have low correlations then the consequences are as follows.

(1) We have two explanatory variables which are highly correlated. Thus, they do not provide as much explanatory power as two variables which have low correlation. We may have wasted time and money collecting two sets of information when one of them alone would have done.
(2) Many of the results of the regression analysis are biased.
(3) The parameters estimates may change quite dramatically on the introduction of a variable which is highly correlated with a variable already in the model.

When two or more explanatory variables in a regression model are highly correlated, we have a case of multicollinearity.

16.2.2 Multiple Tests and Spurious Results

Let us suppose that we have the model above and that we produce five parameter estimates. To test the model, we test if each parameter is significant i.e. H_0: $\alpha = 0$, H_0: $\beta_1 = 0$, ... H_0: $\beta_4 = 0$, and that we carry out these tests at the 10% significance level. What is going to happen?

Firstly, the tests are not independent of each other. They are carried out on the same data and use the same estimate of the variance and the same summary statistics for the data e.g. Σx_i, Σx_i^2. Thus, if we have an odd point in the data set which causes us to overestimate the variance, the variance is overestimated by the same amount for each of the tests. One way to avoid this is to split the data set up into five pieces and to use one piece for each test. This, however, is laborious, impractical without a very large data set, and has problems of its own. A second way around this problem is to be aware of the consequences and not to place too much weight on the results of two or three dependent tests by treating them as independent pieces of evidence.

Secondly, and probably more importantly, we may get a spurious result owing to the multiplicity of the tests. For example, let us assume, for the moment, that the parameters are, in fact, all insignificant i.e. the null hypotheses are true. The framework of hypothesis testing states that, at the 10% significance level, we may reject H_0 in 10% of the cases when it is true. If, for the sake of the probability calculation, we consider the tests to be independent, the probability of failing to reject all five null hypotheses when they are true is $(0.9)^5 = 0.59$. Thus, there is a 0.41 chance of including that at least one parameter is significant. (We must realize that the problem is aggravated as we do not know which one of the five may be expected to yield this false conclusion.) If we carried out the tests at the 5% significance level, this probability is about 0.23 while, for a series of tests at the 20% significance level, the probability is as high as 0.67. If we carry out more than five tests, these probabilities will increase. Thus, by carrying out enough hypothesis tests, we shall eventually conclude that some parameter is significant.

Unfortunately for us, this effect is caused not by the data but by the framework and methodology of hypothesis testing and there is little we can do except to be careful about performing multiple hypothesis tests on a set of data.

16.2.3 Selection of Variables

Which explanatory variables do we use in a multiple regression analysis? The answer to this question has taxed applied statisticians, economists, and business analysts for many years. The ideal model and ideal selection of variables will include most of the following features. We need the explanatory variables to be such that the assumptions of model are approximately valid. We need the variables to be such that the information is cheap to collect. The variables and the structure must make sense. This is particularly important as, no matter what the analyst says a practitioner in another profession is unlikely to believe the analysis if it produces a model totally at odds with his experience. We should not include "too many" variables for an excess of variables will tend to make the structure hazy as well as increasing the likelihood and severity of multicollinearity. We also wish a model which is fairly accurate in fitting the observed values of the response variable.

In summary, we wish a model which is small but adequate, simple yet accurate. Therefore, there is no simple answer to which variables should be included in a model as explanatory variables. All that we can lay down are the guidelines above.

16.3 Regression—Reason for Brevity

The previous chapter and, even more so, this chapter have been relatively short and brief. This is in stark contrast to the number of pages normally devoted to regression in statistics textbooks and to the large number of books which are devoted to regression. However, that is the reason for the brevity here. I feel that it is important to understand the ideas of regression and to have a rough grasp of the technical details. However, more and more, the technicalities are being handled by machines and there is little point in repeating at length what appears in a variety of texts. Therefore, we can leave the large topic of regression now safe in the knowledge not that we have covered the topic completely, but that further information on this topic is readily available in a wide range of textbooks.

CHAPTER 17
Index Numbers

17.1 Introduction

The cost of living rose last year by 5.2%. Obviously, this does not mean that the price of everything we buy rose by exactly 5.2% in the last year. What it means is that the prices of goods and services rose by, on average, 5.2% over the 12-month period. Now, we have already seen that the use of the word "average" can introduce ambiguities. However, in this case, these ambiguities and uncertainties only serve to highlight the problems involved in assessing changes in prices.

Many of the larger building societies regularly issue a House Price Index. One such index is issued by the Nationwide Building Society.

Quarter	Nationwide House Price Index (all properties)	(4th Qtr 1973 = 100)
1973 4th Qtr	100	
1974 4th Qtr	105	
1975 4th Qtr	116	
1976 4th Qtr	125	
1977 4th Qtr	135	
1978 4th Qtr	172	
1979 4th Qtr	225	
1980 4th Qtr	241	
1981 1st Qtr	243	
2nd Qtr	248	
3rd Qtr	248	
4th Qtr	244	
1982 1st Qtr	247	
2nd Qtr	253	
3rd Qtr	256	
4th Qtr	262	
1983 1st Qtr	268	

This type of series provides a quick summary and the numbers are known as Index Numbers; but where do they come from?

In this chapter, we shall be attempting to construct series of index numbers. Many approaches are possible; most of them having both advantages and disadvantages. However, all the series of index numbers will allow us to compare, for example, prices at a particular point in time with prices at some reference point in time. This reference point is generally known as a base.

17.2   Simple Index Numbers for Simple Cases

In U.K. Housing and Construction Statistics, one set of data pertains to Rate Rebates. The average annual rebate is quoted in £'s for 1971 to 1981 and appears below

Year Ending March	Annual Rate Rebate (£)
1971	18.39
1972	22.79
1973	25.77
1974	30.77
1975	36.50
1976	45.50
1977	47.42
1978	51.95
1979	55.80
1980	66.00
1981	79.00

How can we produce a series of index numbers which provides a quick and simple comparison of rate rebates? One way is to fix any year as a base and compare all rate rebates with the rate rebate in this base year, quoting the figures as a percentage.

Thus, if we fix 1971 as a base, we quote all the other years' figures as a percentage relative to the 1971 figures of £18.39. For 1972, we divide 22.79 by 18.39 and multiply to 100 (to convert to a percentage). Most government statistics quote this figure to one decimal place and, in general, we shall follow that example.

Thus, for 1971, the index number is 100. For 1972, it is $22.79/18.39 \times 100 = 123.9$ rounded to one decimal place, and so on for the other years to give

1971	100.0	(1971 = 100)
1972	123.9	
1973	140.1	
1974	167.3	
1975	198.5	
1976	247.4	

1977	257.9
1978	282.5
1979	303.4
1980	358.9
1981	429.6

Thus, we can easily see that there has been a 40.1% increase between 1971 and 1973 and a 203.4% increase between 1971 and 1979.

Note that there has *not* been a 58.4% increase between 1973 and 1975.

The $1971 = 100$ is necessary to clarify which year is the base year. It may have been that the index number for 1973, say, was 100. Then, it would not have been clear which was the base year; this lack of clarity can have serious repercussions as we shall see later.

The choice of base year is sometimes arbitrary and, if we had 1973 as base year, the following series would be obtained.

1971	71.4	$(1973 = 100)$
1972	88.4	
1973	100.0	
1974	119.4	
1975	141.6	
1976	176.6	
1977	184.0	
1978	201.6	
1979	216.5	
1980	256.1	
1981	306.6	

Note that in this simple case, there is a simple mathematical relationship between the two series. We merely multiply each entry in the former series by $18.39/25.77$. In most cases, changing the base year is far more complicated.

Now this simple approach has two problems. Firstly, it only deals with the average. This may sometimes be misleading. Secondly, and more importantly, it only deals with one figure for each year.

Suppose that we have the following data on median purchase prices of flats for a particular area of Southern England.

	Median Price (£)		
Number of Bedrooms	1973	1977	1981
1	6,180	8,450	16,410
2	7,450	10,790	19,820
3	8,190	14,350	23,460
4	12,860	19,520	34,360

How do we construct one series of index numbers for the price of these flats? There are two basic methods; the aggregate method and the relative method.

Aggregate Method

This method involves aggregating the prices for each year. Thus, we imagine that we are buying one of each type of flat. At their median prices, how much would this cost? In 1973, one of each type would cost £34,680. In 1977, the cost would be £53,110 while in 1981, the cost would be £94,050. To get a simple series, we take 1973 as base year and convert the other two figures to percentages relative to £34,680. Thus, we get

Flat Prices		(1973 = 100)
1973	100.0	
1977	153.1	
1981	271.2	

Relative Method

This method involves deriving a series of index numbers for each commodity i.e. for each type of flat and then reducing the four series into one series.

Number of Bedrooms	**1973**	**1977**	**1981**
	(1973 = 100)		
1	100	136.7	265.5
2	100	144.8	266.0
3	100	175.2	286.4
4	100	151.8	267.2

In the simple approach, we merely take the mean of the four index numbers for each year. Thus, we get

Flat Prices		(1973 = 100)
1973	100.0	
1977	152.1	
1981	271.3	

Thus, we have taken these data and, using two different methods, have constructed two series of index numbers. Let us compare and contrast these two methods. Firstly, they are trying to measure or to record changes in prices of flats. However, they are not trying to estimate anything specific as we cannot really identify "the correct answer". Having said that, we can identify procedures which lead to series which are representative of the fluctuations in prices.

The series above are similar. However, the difference in 1977 of 1 unit i.e. of 1% of the 1973 prices could amount to hundreds of thousands of pounds depending upon the size of the market. Also, their similarity does not guarantee that they are representative of the changes in prices.

The aggregate method is simpler and quicker to use in calculation. However, as we have seen in many other areas, this often leads to a loss of information. The relative method is more complex to calculate but provides information on the four types of flats individually as well as collectively.

However, both the methods omit one vital factor: the number of flats of each type which are sold.

17.3 Index Numbers for Different-sized Bundles of Commodities

Let us consider an enlightening but very artificial modification to the previous problem. Suppose that the previous case dealt with 100 flats, each of which was sold in each of the three years, 1973, 1977, 1981. Further, suppose that of these 100 flats, 15 were 1-bedroom, 55 were 2-bedroom, 27 were 3-bedroom and 3 were 4-bedroom flats.

Now, how do we construct a series of index numbers? Well, whatever method we use ought to give greater weight, i.e. greater importance, to the 2-bedroom flats as there are more of them. Also, perhaps 4-bedroom flats ought to have greater weight as a 5% change in the price of one of these has a bigger effect than a 5% change in the price of a cheaper flat. Once again, we shall consider two basic methods.

Aggregate Method

In effect, we consider the hundred flats as a portfolio and we compare the price we would have had to pay for this portfolio were we to buy it in each of the 3 years.

In 1973, the 100 flats would have cost £$(15 \times 6,180 + 55 \times 7,450 + 27 \times 8,190 + 3 \times 12,860)$ which equals £762,160. In 1977, the portfolio would cost £1,157,210. In 1981, the flats cost £2,072,750.

Once again, to get index numbers, we compare with the base year of 1973 to get

1973	100.0	(1973 = 100)
1977	151.8	
1981	272.0	

Commodity	(1) Weight	(2) 1973	(3) 1977	(4) 1981	(5) (1) × (2)	(6) (1) × (3)	(7) (1) × (4)
1–b flat	155,200	100	136.7	265.5	15,520,000	21,215,840	41,205,600
2–b flat	697,767	100	144.8	266.0	69,776,700	101,036,662	185,606,022
3–b flat	414,000	100	175.2	286.4	41,400,000	72,532,800	118,569,600
4–b flat	66,740	100	151.8	267.2	6,674,000	10,121,132	17,832,928

Relative Method

Here, we use the previous relative method construction but introduce weights. For 1-bedroom flats, for example, the weight should reflect (i) the "average" price of the 1-bedroom flats over the time period, and (ii) the number of 1-bedroom flats that there are.

For this period, the "average" price of a 1-bedroom flat will be represented by the mean of the three available figures. This is $(6,180 + 8,450 + 16,410)/3$ which equals 10,346.60.

To get the relative weight for 1-bedroom flats, we multiply this "average" price by the number of these flats, i.e. by 15 to get 155,200.

For 2-bedroom flats, the weight is $55 \times (7,450 + 10,790 + 19,820)/3$ which equals 697,766.6 rounded up to 697,767.

For 3-bedroom flats, the weight is $27 \times (8,190 + 14,350 + 23,460)/3$ which equals 414,000, and for 4-bedroom flats, the weight is 66,740.

Now, in the previous use of the relative method, for 1977, we merely took the mean of the four simple index numbers and got 152.1. Now, however, we are going to take a weighted mean.

Perhaps, the simplest way is to do this in a table.

Now, if we sum up column (5), we get 133,370,700. Thus, figure does not have any units which can be easily interpreted but it is the weighted sum for 1973. (Above we talked about weighted means but that only involves dividing by 100. As we are going to divide one weighted sum by another, dividing top and bottom by 100 is a waste of time).

If we sum column (6), we get 204,916,434, the weighted sum for 1977, while the sum of column (7) is 363,214,150, the weighted sum for 1981.

Once again, if we compare these to get a series of index numbers, we get

1973	100.0	(1973 = 100)
1977	153.6	
1981	272.3	

So far, we have produced four different series of Index Numbers for the same data.

(1973 = 100)	**1973**	**1977**	**1981**
Aggregate Method (Simple)	100	153.1	271.2
Relative Method (Simple)	100	152.1	271.3
Aggregate Method (with weights)	100	151.8	272.0
Relative Method (with weights)	100	153.6	272.3

Again, it ought to be noted that while these figures look similar, a difference of 1% in total 1973 prices is about £8,000—possibly the price of a 2-bedroom or 3-bedroom flat.

However, more importantly, there is still one other major problem. In almost all cases, the number of each commodity sold is unlikely to remain the same from year to year. This is due to several reasons including human behaviour patterns, the uncontrollability of the situation, changes in population, changes in actual demand, and the inherent problems of accuracy of data. Another reason is the basic feature of a price-consumption relationship. We are trying to measure changes in price yet we know that changes in price will affect consumption.

There are, of course, several other reasons why sales figures or consumption figures may change from year to year. We shall not discuss these reasons further but, in the next section, we shall discuss methods of coping with these changing figures.

17.4 A General Approach to Price Indices

In this section, we shall discuss the general problems of constructing series of index numbers. As a simple illustrative example, we can consider Joe Honest, a car salesman.

If we only consider three models of car, we have the following data:

	1974		1975		1976	
Model	Number Sold	Price	Number Sold	Price	Number Sold	Price
A	28	£1,500	35	£1,600	49	£2,000
B	6	£3,000	6	£3,500	9	£4,000
C	2	£5,500	3	£6,500	4	£8,000

How do we construct an index of prices? We can merely look at prices but clearly the quantity sold must enter into our consideration. To take the matter to an extreme, we may have a particular model of car which cost £10,000 but none of them were sold. This price of £10,000 should not be included or, technically speaking, it should be included with a relative weight of zero.

Thus, we must introduce weights where a weight represents the importance of a commodity's prices relative to the importance of the other commodities' prices.

17.4.1 Aggregate Method—Base-Year Weighted

Here, we shall extend the aggregate method and will derive relative weights from the base year which, in this case, is 1974. If we have a price index, the only other piece of information available to use as a relative weight is the number sold.

In effect, what we are to do is to compare the prices in different years relative to the quantities sold in the base year of 1974. We are using the aggregate approach and so we are interested in total revenue. For 1974, the revenue is $(28 \times 1,500) + (6 \times 3,000) + (2 \times 5,500)$ which equals £71,000. For 1975, we are interested in the revenue generated from selling the 1974 quantities at 1975 prices. This revenue is $(28 \times 1,600) + (6 \times 3,500) + (2 \times 6,500) =$ £78,800. Finally, for 1976, we are interested in the revenue generated from selling the 1974 quantities at 1976 prices. This equals $(28 \times 2,000) + (6 \times 4,000) + (2 \times 8,000) =$ £96,000.

What we have done is to keep the quantities sold the same and so we are comparing prices with prices. By dividing each of the three figures by the 1974 figure, and multiplying by 100, we get the following series of index numbers.

1974	100.0	1974 = 100
1975	111.0	
1976	135.2	

This index has many appealing qualities. It is simple to calculate and makes intuitive sense as we are comparing all sets of prices relative to the quantities sold in the base year. This set of commodities has various terms in economics including a commodity bundle, a basket of goods, and a portfolio. The Retail Price Index (RPI) has such a basket of goods and the total cost of the basket is monitored from month to month.

17.4.2   Relative Method—Base-Year Weighted

In a similar way to the previous section, this is also an extension of the previous occurrence of the relative method. Probably the best way to examine this method is in the form of a table.

In column (1), we have the weight, i.e. the quantities sold. Columns (2) to (4) contain the prices for the commodities for the three years of interest. In columns (5) to (7) appear the three relative series as in the simpler approach in 17.2. However, instead of taking a simple arithmetic mean of each of the columns, we need a weighted mean, with each component having importance relative to the numbers sold. We are going to compare these weighted means and so we only need weighted sums. The next three columns arise from multiplying the elements in column (1) by those in (5), (6) and (7), respectively. We add up these last

3 columns and compare them to get the series of index numbers. For 1975, the index number is 3,923.14/3,600 × 100 and for 1976, it is 4,824.12/3,600 × 100. Thus, we have

1974	100.0	1974 = 100
1975	109.0	
1976	134.0	

This index is not as simple as the preceding index to construct but does, once again, provide valuable information as an extra. In fact, there is a simpler arithmetic way to construct the index but it loses the extra information and it is more difficult to see exactly what is going on.

These two indices are base-year weighted, i.e. the base-year provides the relative weight. This idea was first introduced by an economist called Etienne Laspeyres. The advantage is that several years of results can be compared without the need to evaluate a new set of weights. The disadvantage is that, as we are dealing with economics and, often, with human behaviour, these weights very soon become out of date. For example, in the RPI, in the late 1950s, the relative weight on food prices was about 35%. By 1966 this weight was under 30%. Over the same interval of time, the relative weighting for transport rose from 6.8% to almost 12%. As a result, H. Paasche, another economist, introduced the idea of current-year weighting—using the current year to provide weights.

17.4.3 Aggregate Method—Current-Year Weighted

Instead of using the base year to provide weights, we use the current year. Thus, if we are trying to evaluate an index number for 1975, we use the 1975 figures to get a relative weighting. If we are looking at 1976, we use 1976 figures to get a relative weighting.

For 1974, the index number will be 100, as 1974 is the base year. For 1975, in the base-year weighted approach in 17.4.2, we use the base year commodities as the weights and compared 1974 commodities at 1975 prices with 1974 commodities at 1974 prices. Now, we are using 1975 figures to provide the weights and so we compare 1975 commodities at 1975 prices with 1975 commodities at 1974 prices. Once again, we are comparing sets of prices keeping the commodity bundle constant.

In this case, the revenue from 1975 commodities at 1975 prices is £96,500 while the revenue from 1975 commodities at 1974 prices is £87,000. Thus, the index number for 1975 is 96,500/87,000 × 100 = 110.9.

Model	Number Sold 1974 (1)	Price 1974 (2)	Price 1975 (3)	Price 1976 (4)	Relative Price (5)	(6)	(7)	(1) × (5)	(1) × (6)	(1) × (7)
A	28	1,500	1,600	2,000	100	106.67	133.33	2,800	2,986.76	3,733.24
B	6	3,000	3,500	4,000	100	116.67	133.33	600	700.02	799.98
C	2	5,500	6,500	8,000	100	118.18	145.45	200	236.36	290.90
								3,600	3,923.14	4,824.12

For 1976, the revenue from 1976 commodities at 1976 prices is £166,000. For 1976 commodities at 1974 prices the revenue is £122,500. The index number for 1976 is, therefore, $166,000/122,500 \times 100 = 135.5$.

Thus, the index series is

1974	100.0	$1974 = 100$
1975	110.9	
1976	135.5	

17.4.4 Relative Method—Current-Year Weighted

Once again, we take the basic relative method and extend it. As in 17.4.4, we derive the relative weightings from the current year and not from the base year. A table will clarify the calculations and manipulations.

For the index for 1975, we need to compare 1975 commodities at 1975 relative prices with 1975 commodities at 1974 relative prices. The former are to be found by multiplying the elements in columns (2) and (8) to give column (13). The latter are in column (11). If we sum these and compare, the index number is

$$\frac{4,788.01}{4,400} \times 100 = 108.8$$

For 1976, we need to compare 1976 commodities at 1976 relative prices (column (14)) with 1976 commodities at 1974 relative prices (column (12)). The resultant index number is $7,784.94/6,200 \times 100 = 125.6$.

In fact, columns (10), (11) and (12) are superfluous here as there are merely one hundred times the entries in columns (1), (2) and (3). As a result, we needed only to compare column (13) with column (2) and column (14) with column (3).

17.5 A Better Price Index

We have now constructed four more series of index numbers.

1974 = 100	1973	1975	1976
Aggregate (base-year weighted)	100	110.0	135.2
Relative (base-year weighted)	100	109.0	134.0
Aggregate (current-year weighted)	100	110.9	135.5
Relative (current-year weighted)	100	108.8	125.6

Which one is correct? Well, in fact, it is impossible to say that

Model	Number Sold			Price			Relative Price			(10)	(11)	(12)	(13)	(14)
	1974	1975	1976	1974	1975	1976				(1) × (7)	(2) × (7)	(3) × (7)	(2) × (8)	(3) × (9)
	(1)	(2)	(3)	(4)	(5)	(6)	(7)	(8)	(9)					
A	28	35	49	1,500	1,600	2,000	100	106.67	133.33	2,800	3,500	4,900	3,733.45	6,533.17
B	6	6	9	3,000	3,500	4,000	100	116.67	133.33	600	600	900	700.02	1,199.97
C	2	3	4	5,500	6,500	8,000	100	118.18	145.45	200	300	400	354.54	581.80
	36	44	62							3,600	4,400	6,200	4,788.01	7,784.94

one method is correct and the others wrong. They are all attempting to reflect changes in a market by measuring separate commodities. They each have their own problems and the fact that they give similar results does not necessarily imply that they are a close reflection of the state of the market.

The search for an improvement to these indices has engaged economists and econometricians for many years. Some people have proposed better indices while others have tried to define what constitutes a "perfect" index.

Many proposals have revolved around the idea of taking some sort of average. (This is only valid if we can assume that the index numbers of all the series are close to an accurate reflection of the market). Frank Edgeworth proposed taking the arithmetic mean of the weights while I. Fisher proposed taking the geometric mean of the index numbers for each year. (Recall that, for two numbers, the geometric mean is the square root of the product, for three numbers, the cube root of the product, . . .).

One of the indices used by the *Financial Times* uses a geometric mean. However, this idea has not encountered much enthusiasm elsewhere.

At this point we could discuss an ideal index but, first of all, we must look at the idea of a quantity index.

17.6   A Quantity Index

So far, we have been constructing price indices, i.e. a series of index numbers which reflect the changes in prices. However, we are often interested in quantity indices to reflect changes in the level of consumption or the level of production or the level of sales.

The techniques for constructing quantity indices are similar to those for price indices. The only difference is that we use the information on prices to provide the relative weightings for comparing quantities. In other words, the rôles of price data and quantity are reversed. There is no need to go through all the manipulations again; one example should be sufficient.

For the car salesman, we shall construct a current-year weighted quantity index. This should reflect the variation in level of sales. Further, we shall use the aggregate method. This means that we use the current year prices as relative weights.

For 1974, the index number is 100. For 1975, we compare 1975 quantities at 1975 prices with 1974 quantities at 1975 prices. 1975 quantities at 1975 prices generate revenue of £96,500. 1974 quantities at 1975 prices generate revenue of £78,800. Thus, the index number for 1975 is $99,500/78,800 \times 100 = 122.5$.

For 1976, the index number is $166,000/96,000 \times 100 = 172.9$.

If we once again, calculate four different quantity indices, we get

1974 = 100	1973	1975	1976
Aggregate (base-year weighted)	100	122.5	172.5
Relative (base-year weighted)	100	131.3	181.3
Aggregate (current-year weighted)	100	122.5	172.9
Relative (current-year weighted)	100	131.5	182.1

17.7 A Perfect Index

Several properties have been proposed and discussed as properties which a perfect index should have. Two which we shall briefly mention are circularity and time reversal.

17.7.1 Circularity

In effect, circularity really means that the index is transitive and that the calculations carry across years. To consider a specific example, we can look at the current year weighted aggregate quantity index constructed above.

Using 1974 as a base year, the index for 1975 was 122.5. If we used 1975 as a base year, the index for 1976 would have been 142.4. If the index had the circularity property then the index for 1976 using 1974 as a base year would be $1.424 \times 1.225 \times 100 = 174.4$. It does not and, therefore, this index does not have the circularity property.

If the circularity property held, this would mean that the introduction of one or more intermediate base years has no effect on a comparison.

17.7.2 Time Reversal

In effect, this property states that the index for each of two points in time bear the same inter-relationship irrespective of which one is used as a base year.

Considering again the above example of a current year weighted aggregate quantity index, if 1974 is the base year, the index for 1976 is 172.9. If 1976 is the base year, then the index for 1974 can be calculated to be 57.8.

In this case, if 1974 is the base year, the index for 1976 divided by the index for 1974 is $172.9/100 = 1.729$. If 1976 is the base year, the index for 1976 divided by the index for 1974 is $100/57.8 = 1.730$. These are, in fact, the same, except for rounding error as both of them arise from dividing 166 by 96.

17.8   Some other Topics

Index numbers are used in practice and, as a result, theoretical considerations, alone, are not sufficient to use them properly. Two practical issues will be briefly dealt with here.

17.8.1   Changing the Base

From time to time, it becomes advisable to change the base year. This may be so because the base year was long ago or exhibited very different characteristics to the present time period or merely to enable users to shorten the length of series quoted. In certain types of indices, changing the base is simply a matter of arithmetic, i.e. we rescale by dividing all the index numbers by the index number for the new base year and multiply by 100.

For example, suppose that we had a set of index numbers such as

1971	100.0	(1971 = 100)
1972	104.1	
1973	104.3	
1974	117.6	
1975	123.2	
1976	125.1	
1977	128.3	
1978	141.6	

We could change the base year to 1975 by dividing all entries by 123.2 and multiplying by 100.

Thus we now have

1971	81.2	(1975 = 100)
1972	84.5	
1973	88.7	
1974	95.5	
1975	100.0	
1976	101.5	
1977	104.1	
1978	114.9	

This simple method is only valid mathematically in the case of simple aggregate indices such as those discussed in 17.2 and 17.3.

For more complex (and more accurate) indices, the mathematics is invalid. In these cases, it is necessary to recalculate the index using the new base year.

17.8.2 Comparison of Two Indices

Suppose that two factories have produced indices of productivity.

(1975 = 100)

Factory A	97.5	100.0	102.7	104.3	106.6	109.2	113.5	115.8
	1974	1975	1976	1877	1978	1979	1980	1981
Factory B	96.6	100.0	103.1	105.6	108.4	110.0	116.2	119.3

Under what conditions might we conclude that Factory B has higher productivity than Factor A? Well, clearly the indices must be on comparable scales, i.e. did the factories have the same productivity in 1975? Are the weights and type of index used the same? If so, then, subject to some minor conditions, we may wish to conclude that Factory B has higher productivity. If not, then probably the only conclusion we can reach is that Factory B is increasing its productivity at a faster rate than Factory A.

Examples

1. In a sector of the UK, personal savings can be thought of as being divided into 4 areas; National Savings (administered by the government), Building Society deposits, Life Assurance and Superannuation Fund deposits, and Bank deposits. The nett deposits (in £m) for each of these areas for the last 4 years is as follows:

	National Savings	Building Societies	Life Assurance	Banks
1981	177	102	610	18
1982	166	118	628	11
1983	141	126	631	11
1984	153	134	685	9

Construct a suitable index to record the change in this sector's behaviour as regards personal savings. What other calculations would be important if you, as an employee of a building society (or of the Building Societies Association), wish to ascertain how your share of the market is faring?

2. The 5 set books for a course in building technology did not

change from 1976 to 1982. The total price of the 5 books in these 7 years was

1976	1977	1978	1979	1980	1981	1982
£11.35	£12.85	£15.00	£15.75	£18.20	£20.15	£23.90

Construct an index for this set of prices.

3. A College of further education has 4 grades of academic staff. For the last 5 years, we have information on the number of employees in each grade and their median salary (in £000's).

	Head of Department		Principal Lecturer		Senior Lecturer		Lecturer	
1979	8	16.4	25	13.6	58	11.8	110	10.2
1980	8	16.8	28	13.9	64	12.3	116	10.8
1981	9	17.5	29	14.1	68	12.9	123	11.3
1982	11	18.5	30	14.3	67	13.3	127	11.7
1983	11	19.7	32	14.7	65	13.6	128	12.1

Construct an appropriate index to reflect the changes in the college's total wage bill. Construct another index which shows how the salaries paid to employees have varied over 5 years. A third index may be constructed to reflect how the number of staff has varied over 5 years. Construct this and suggest some economic problems with its interpretation.

4. The Smallville Tenants' Association requires information about the relative increases in the rents paid to the regional housing department. The data collected are as follows:
Weekly median rents (£)

	1968		1979		1980	
Type of Housing	Number of Tenancies	Rent	Number	Rent	Number	Rent
A	100	4.0	100	9.0	100	22.0
B	44	4.0	80	7.5	75	20.0
C	27	4.2	116	8.0	160	18.5
D	4	5.0	20	10.0	70	25.0

Calculate an index to assist the Tenants' Association. State clearly what its limitations are. Also, calculate an index which may be of greater use to the region's finance director as it will give him/her information on rental income.

5. A small engineering company in Lancashire has the following profile of staff and mean weekly salaries (in £)

Type of	1965		1971		1977	
		Weekly				
Worker	Number	Salary	Number	Salary	Number	Salary
Unskilled	9	16.5	10	28.5	10	53.4
Semi-skilled	23	19.0	28	36.2	27	65.0
Skilled	17	24.0	19	49.3	29	80.0
Clerical	1	17.4	2	18.0	4	61.2

The managing director is interested in relative increases in the employees' salaries. Calculate an appropriate index to assist him and advise him as to the index's underlying assumptions and limitations.

Tables

Table 1 Normal Probability Distribution

The function tabulated is $1 - \Phi(z)$ is the cumulative distribution function of a standardized Normal random variable Z. In other words, the tabulated entries are the probabilities that a standardized Normal random variable Z exceeds some value z: $P(Z > z)$

	.00	.01	.02	.03	.04	.05	.06	.07	.08	.09
0.0	.5000	.4960	.4920	.4880	.4840	.4801	.4761	.4721	.4681	.4641
0.1	.4602	.4562	.4522	.4483	.4443	.4404	.4364	.4325	.4286	.4247
0.2	.4207	.4168	.4129	.4090	.4052	.4013	.3874	.3936	.3897	.3859
0.3	.3821	.3783	.3745	.3707	.3669	.3632	.3594	.3557	.3520	.3482
0.4	.3446	.3409	.3372	.3336	.3300	.3264	.3228	.3192	.3156	.3121
0.5	.3085	.3050	.3015	.2981	.2946	.2912	.2877	.2843	.2810	.2776
0.6	.2743	.2709	.2676	.2643	.2611	.2578	.2546	.2514	.2483	.2451
0.7	.2420	.2389	.2358	.2327	.2296	.2266	.2236	.2206	.2177	.2148
0.8	.2119	.2090	.2061	.2033	.2005	.1977	.1949	.1922	.1894	.1867
0.9	.1841	.1814	.1788	.1762	.1736	.1711	.1685	.1660	.1635	.1611
1.0	.1587	.1562	.1539	.1515	.1492	.1469	.1446	.1423	.1401	.1379
1.1	.1357	.1335	.1314	.1292	.1271	.1251	.1230	.1210	.1190	.1170
1.2	.1151	.1131	.1112	.1093	.1075	.1056	.1038	.1020	.1003	.0985
1.3	.0968	.0951	.0934	.0918	.0901	.0885	.0869	.0853	.0838	.0823
1.4	.0808	.0793	.0778	.0764	.0749	.0735	.0721	.0708	.0694	.0681
1.5	.0668	.0665	.0643	.0630	.0618	.0606	.0594	.0582	.0571	.0559
1.6	.0548	.0537	.0526	.0516	.0505	.0495	.0485	.0475	.0465	.0445
1.7	.0446	.0436	.0427	.0418	.0409	.0401	.0392	.0384	.0375	.0367
1.8	.0359	.0351	.0344	.0336	.0329	.0322	.0314	.0307	.0301	.0294
1.9	.0287	.0281	.0274	.0268	.0262	.0256	.0250	.0244	.0239	.0233
2.0	.02275	.02222	.02169	.02118	.02068	.02018	.01970	.01923	.01876	.01831
2.1	.01786	.01743	.01700	.01659	.01618	.01578	.01539	.01500	.01463	.01426
2.2	.01390	.01355	.02321	.01287	.01255	.01222	.01191	.01160	.01130	.01101
2.3	.01072	.01044	.01017	.00990	.00964	.00939	.00914	.00889	.00866	.00842
2.4	.00820	.00798	.00776	.00755	.00734	.00714	.00695	.00676	.00657	.00639
2.5	.00621	.00604	.00587	.00570	.00554	.00539	.00523	.00508	.00494	.00480
2.6	.00466	.00453	.00440	.00427	.00415	.00402	.00391	.00379	.00368	.00357
2.7	.00347	.00336	.00326	.00317	.00307	.00298	.00289	.00280	.00272	.00264
2.8	.00256	.00248	.00240	.00233	.00226	.00219	.00212	.00205	.00199	.00193
2.9	.00187	.00181	.00175	.00169	.00164	.00159	.00154	.00149	.00144	.00139
3.0	.00135					.00114				
3.1	.00097					.00082				
3.2	.00069					.00058				
3.3	.00048					.00040				
3.4	.00034					.00028				
3.5	.00023					.00019				
3.6	.00016					.00013				
3.7	.00011					.00009				
3.8	.00007					.00006				
3.9	.00005					.00004				
4.0	.00003									

Tables 1–3 are taken from Tables IIi, IV, and V of Fisher & Yates: Statistical Tables for Biological, Agricultural and Medical Research published by Longman Group Ltd, London (previously published by Oliver and Boyd Ltd, Edinburgh) and by permission of the authors and publishers.

Table 2 Percentage points of the χ^2

The function tabulated is $\chi^2_{\nu\alpha}$: the 100α percentage point of the χ^2 distribution with degrees of freedom. In other words, $P(\chi^2_\nu > \chi^2_{\nu\alpha}) = \alpha$

$\sigma=$	.995	.99	.98	.975	.95	.90	.80	.75	.70	.50
$\nu=1$	$.0^4393$	$.0^3157$	$.0^3628$	$.0^3982$	.00393	.0158	.0642	.102	.148	.455
2	.0100	.0201	.0404	.0506	.103	.211	.446	.575	.713	1.386
3	.0717	.115	.185	.216	.352	.584	1.005	1.213	1.424	2.366
4	.207	.297	.429	.484	.711	1.064	1.649	1.923	2.195	3.357
5	.412	.554	.752	.831	1.145	1.610	2.343	2.675	3.000	4.351
6	.676	.872	1.134	1.237	1.635	2.204	3.070	3.455	3.828	5.348
7	.989	1.239	1.564	1.690	2.167	2.833	3.822	4.255	4.671	6.346
8	1.344	1.646	2.032	2.180	2.733	3.490	4.594	5.071	5.527	7.344
9	1.735	2.088	2.532	2.700	3.325	4.168	5.380	5.899	6.393	8.343
10	2.156	2.558	3.059	3.247	3.940	4.865	6.179	6.737	7.267	9.342
11	2.603	3.053	3.609	3.816	4.575	5.578	6.989	7.584	8.148	10.341
12	3.074	3.571	4.178	4.404	5.226	6.304	7.807	8.438	9.034	11.340
13	3.565	4.107	4.765	5.009	5.892	7.042	8.634	9.299	9.926	12.340
14	4.075	4.660	5.368	5.629	6.571	7.790	9.467	10.165	10.821	13.339
15	4.601	5.229	5.985	6.262	7.261	8.547	10.307	11.036	11.721	14.339
16	5.142	5.812	6.614	6.908	7.962	9.312	11.152	11.912	12.624	15.338
17	5.697	6.408	7.255	7.564	8.672	10.085	12.002	12.792	13.531	16.338
18	6.265	7.015	7.906	8.231	9.390	10.865	12.857	13.675	14.440	17.338
19	6.844	7.633	8.567	8.907	10.117	11.651	13.716	14.562	15.352	18.338
20	7.434	8.260	9.237	9.591	10.851	12.443	14.578	15.452	16.266	19.337
21	8.034	8.897	9.915	10.283	11.591	13.240	15.445	16.344	17.182	20.337
22	8.643	9.542	10.600	10.982	12.338	14.041	16.314	17.240	18.101	21.337
23	9.260	10.196	11.293	11.688	13.091	14.848	17.187	18.137	19.021	22.337
24	9.886	10.856	11.992	12.401	13.848	15.659	18.062	19.037	19.943	23.337
25	10.520	11.524	12.697	13.120	14.611	16.473	18.940	19.039	20.867	24.337
26	11.160	12.198	13.409	13.844	15.379	17.292	19.820	20.843	21.792	25.336
27	11.808	12.879	14.125	14.573	16.151	18.114	20.703	21.749	22.719	26.336
28	12.461	13.565	14.847	15.308	16.928	18.939	21.588	22.657	23.647	27.336
29	13.121	14.256	15.574	16.047	17.708	19.768	22.475	23.567	24.577	28.336
30	13.787	14.953	16.306	16.791	18.493	20.599	23.364	24.478	25.508	29.336
40	20.706	22.164	23.838	24.433	26.509	29.051	32.345	33.660	34.872	39.335
50	27.991	29.707	31.664	32.357	34.764	37.689	41.449	42.942	44.313	49.335
60	35.535	37.485	39.699	40.482	43.188	46.459	50.641	52.294	53.809	59.335
70	43.275	45.442	47.893	48.758	51.739	55.329	59.898	61.698	63.346	69.334
80	51.171	53.539	56.213	57.153	60.391	64.278	69.207	71.145	72.915	79.334
90	59.196	61.754	64.634	65.646	69.126	73.291	78.558	80.625	82.511	89.334
100	67.327	70.065	73.142	74.222	77.929	82.358	87.945	90.133	92.129	99.334

For very large values of ν, approximate values can be obtained by using the fact that $\sqrt{2\chi^2}$ has an approximate Normal distribution with mean $\sqrt{2\nu-1}$ and a variance of 1.

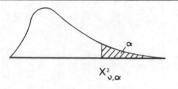

$X^2_{\nu,\alpha}$

.30	.25	.20	.10	.05	.025	.02	.01	.005	.001	$= \alpha$
1.074	1.323	1.642	2.706	3.841	5.024	5.412	6.635	7.879	10.827	$\nu = 1$
2.408	2.773	3.219	4.605	5.991	7.378	7.824	9.210	10.597	13.815	2
3.665	4.108	4.642	6.251	7.815	9.348	9.837	11.345	12.838	16.268	3
4.878	5.385	5.989	7.779	9.488	11.143	11.668	13.277	14.860	18.465	4
6.064	6.626	7.289	9.236	11.070	12.832	13.388	15.086	16.750	20.517	5
7.231	7.841	8.558	10.645	12.592	14.449	15.033	16.812	18.548	22.457	6
8.383	9.037	9.803	12.017	14.067	16.013	16.622	18.475	20.278	24.322	7
9.524	10.219	11.030	13.362	15.507	17.535	18.168	20.090	21.955	26.125	8
10.656	11.389	12.242	14.684	16.919	19.023	19.679	21.666	23.589	27.877	9
11.781	12.549	13.442	15.987	18.307	20.483	21.161	23.209	25.188	29.588	10
12.899	13.701	14.631	17.275	19.675	21.920	22.618	24.725	26.757	31.264	11
14.011	14.845	15.812	18.549	21.026	23.337	24.054	26.217	28.300	32.909	12
15.119	15.984	16.985	19.812	22.362	24.736	25.472	27.688	29.819	34.528	13
16.222	17.117	18.151	21.064	23.685	26.119	26.873	29.141	31.319	36.123	14
17.322	18.245	19.311	22.307	24.996	27.488	28.259	30.578	32.801	37.697	15
18.418	19.369	20.465	23.542	26.296	28.845	29.633	32.000	34.267	39.252	16
19.511	20.489	21.615	24.769	27.587	30.191	30.995	33.409	35.718	40.790	17
20.601	21.605	22.760	25.989	28.869	31.526	32.346	34.805	37.156	42.312	18
21.689	22.718	23.900	27.204	30.144	32.852	33.687	36.191	38.582	43.820	19
22.775	23.828	25.028	28.412	31.410	34.170	35.020	37.566	39.997	45.315	20
23.858	24.935	26.171	29.615	32.671	35.479	36.343	38.932	41.401	46.797	21
24.939	26.039	27.301	30.813	33.924	36.781	37.659	40.289	42.796	48.268	22
26.018	27.141	28.429	32.007	35.172	38.076	38.968	41.638	44.181	49.728	23
27.096	28.241	29.553	33.196	36.415	39.364	40.270	42.980	45.558	51.179	24
28.172	29.339	30.675	34.382	37.652	40.646	41.566	44.314	46.928	52.620	25
29.246	30.434	31.795	35.563	38.885	41.923	42.856	45.642	48.290	54.052	26
30.319	31.528	32.912	36.741	40.113	43.194	44.140	46.963	49.645	55.476	27
31.391	32.620	34.027	37.916	41.337	44.461	45.419	48.278	50.993	56.893	28
32.461	33.711	35.139	39.087	42.557	45.722	46.693	49.588	52.336	58.302	29
33.530	34.800	36.250	40.256	43.773	46.979	47.962	50.892	53.672	59.703	30
44.165	45.616	47.269	51.805	55.759	59.342	60.436	63.691	66.766	73.402	40
54.723	56.334	58.164	63.167	67.505	71.420	72.613	76.154	79.490	86.661	50
65.227	66.981	68.972	74.397	79.082	83.298	84.580	88.379	91.952	99.607	60
75.689	77.577	79.715	85.527	90.531	95.023	96.388	100.425	104.215	112.317	70
86.120	88.130	90.405	96.578	101.880	106.629	108.069	112.329	116.321	124.839	80
96.524	98.650	101.054	107.565	113.145	118.136	119.648	124.116	128.299	137.208	90
106.906	109.141	111.667	118.498	124.342	129.561	131.142	135.807	140.170	149.449	100

Table 3 Percentage Points of the F Distribution

The table below gives the values of F_{α, ν_1, ν_2}, the 100α percentage point of the F distribution having ν_1 and ν_2 degrees of freedom in the numerator and denominator, respectively. For each pair of values of ν_1 and ν_2, F_{α, ν_1, ν_2} is tabulated for $\alpha = 0.05$, 0.025, 0.01, and 0.001, the values for $\alpha = 0.025$ being bracketed. For the lower percentage points of the distribution, $F_{1-\alpha, \nu_1, \nu_2} = 1/(F_{\alpha, \nu_1, \nu_2})$, e.g.

$$F_{.99, 4, 7} = 1/(F_{.01, 7, 4}) = 1/15.0 = 0.0667.$$

ν_2 \ ν_1	1	2	3	4	5	6	7	8	10	12	24	∞
1	161.4	199.5	215.7	224.6	230.2	234.0	236.8	238.9	241.9	243.9	249.0	254.3
	(648)	(800)	(864)	(900)	(922)	(937)	(948)	(957)	(969)	(977)	(997)	(1,018)
	4,052	5,000	5,403	5,625	5,764	5,859	5,928	5,981	6,056	6,106	6,235	6,366
	*405,200	500,000	540,400	562,500	576,400	585,900	592,900	598,100	605,600	610,700	623,500	636,600
2	18.5	19.0	19.2	19.2	19.3	19.3	19.4	19.4	19.4	19.4	19.5	19.5
	(38.5)	(39.0)	(39.2)	(39.2)	(39.3)	(39.3)	(39.4)	(39.4)	(39.4)	(39.4)	(39.5)	(39.5)
	98.5	99.0	99.2	99.2	99.3	99.3	99.4	99.4	99.4	99.4	99.5	99.5
	998.5	999.0	999.2	999.2	999.3	999.3	999.4	999.4	999.4	999.4	999.5	999.5
3	10.13	9.55	9.28	9.12	9.01	8.94	8.89	8.85	8.79	8.74	8.64	8.53
	(17.4)	(16.0)	(15.4)	(15.1)	(14.9)	(14.7)	(14.6)	(14.5)	(14.4)	(14.3)	(14.1)	(13.9)
	34.1	30.8	29.5	28.7	28.2	27.9	27.7	27.5	27.2	27.1	26.6	26.1
	167.0	148.5	141.1	137.1	134.6	132.8	131.5	130.6	129.2	128.3	125.9	123.5
4	7.71	6.94	6.59	6.39	6.26	6.16	6.09	6.04	5.96	5.91	5.77	5.63
	(12.22)	(10.65)	(9.98)	(9.60)	(9.36)	(9.20)	(9.07)	(8.98)	(8.84)	(8.75)	(8.51)	(8.26)
	21.2	18.0	16.7	16.0	15.5	15.2	15.0	14.8	14.5	14.4	13.9	13.5
	74.14	61.25	56.18	53.44	51.71	50.53	49.66	49.00	48.05	47.41	45.77	44.05
5	6.61	5.79	5.41	5.19	5.05	4.95	4.88	4.82	4.74	4.68	4.53	4.36
	(10.01)	(8.43)	(7.76)	(7.39)	(7.15)	(6.98)	(6.85)	(6.76)	(6.62)	(6.52)	(6.28)	(6.02)
	16.26	13.27	12.06	11.39	10.97	10.67	10.46	10.29	10.05	9.89	9.47	9.02
	47.18	37,12	33.20	31.09	29.75	28.83	28.16	27.65	26.92	26.42	25.14	23.79

F_{α, ν_1, ν_2}

6	5.99	5.14	4.76	4.53	4.39	4.28	4.21	4.15	4.06	4.00	3.84	3.67
	(8.81)	(7.26)	(6.60)	(6.23)	(5.99)	(5.82)	(5.70)	(5.60)	(5.46)	(5.37)	(5.12)	(4.85)
	13.74	10.92	9.78	9.15	8.75	8.47	8.26	8.10	7.87	7.72	7.31	6.88
	35.51	27.00	23.70	21.92	20.80	20.03	19.46	19.03	18.41	17.99	16.90	15.75
7	5.59	4.74	4.35	4.12	3.97	3.87	3.79	3.73	3.64	3.57	3.41	3.23
	(8.07)	(6.54)	(5.89)	(5.52)	(5.29)	(5.12)	(4.99)	(4.90)	(4.76)	(4.67)	(4.42)	(4.14)
	12.25	9.55	8.45	7.85	7.46	7.19	6.99	6.84	6.62	6.47	6.07	5.65
	29.25	21.69	18.77	17.20	16.21	15.52	15.02	14.63	14.08	13.71	12.73	11.70
8	5.32	4.46	4.07	3.84	3.69	3.58	3.50	3.44	3.35	3.28	3.12	2.93
	(7.57)	(6.06)	(5.42)	(5.05)	(4.82)	(4.65)	(4.53)	(4.43)	(4.30)	(4.20)	(3.95)	(3.67)
	11.26	8.65	7.59	7.01	6.63	6.37	6.18	6.03	5.81	5.67	5.28	4.86
	25.42	18.49	15.83	14.39	13.48	12.86	12.40	12.05	11.54	11.19	10.30	9.34
9	5.12	4.26	3.86	3.63	3.48	3.37	3.29	3.23	3.14	3.07	2.90	2.71
	(7.21)	(5.71)	(5.08)	(4.72)	(4.48)	(4.32)	(4.20)	(4.10)	(3.96)	(3.87)	(3.61)	(3.33)
	10.56	8.02	6.99	6.42	6.06	5.80	5.61	5.47	5.26	5.11	4.73	4.31
	22.86	16.39	13.90	12.56	11.71	11.13	10.69	10.37	9.87	9.57	8.72	7.81
10	4.96	4.10	3.71	3.48	3.33	3.22	3.14	3.07	2.98	2.91	2.74	2.54
	(6.94)	(5.46)	(4.83)	(4.47)	(4.24)	(4.07)	(3.95)	(3.85)	(3.72)	(3.62)	(3.37)	(3.08)
	10.04	7.56	6.55	5.99	5.64	5.39	5.20	5.06	4.85	4.71	4.33	3.91
	21.04	14.91	12.55	11.28	10.48	9.93	9.52	9.20	8.74	8.44	7.64	6.76
11	4.84	3.98	3.59	3.36	3.20	3.09	3.01	2.95	2.85	2.79	2.61	2.40
	(6.72)	(5.26)	(4.63)	(4.28)	(4.04)	(3.88)	(3.76)	(3.66)	(3.53)	(3.43)	(3.17)	(2.88)
	9.65	7.21	6.22	5.67	5.32	5.07	4.89	4.74	4.54	4.40	4.02	3.60
	19.69	13.81	11.56	10.35	9.58	9.05	8.66	8.35	7.92	7.63	6.85	6.00
12	4.75	3.89	3.49	3.26	3.11	3.00	2.91	2.85	2.75	2.69	2.51	2.30
	(6.55)	(5.10)	(4.47)	(4.12)	(3.89)	(3.73)	(3.61)	(3.51)	(3.37)	(3.28)	(3.02)	(2.72)
	9.33	6.93	5.95	5.41	5.06	4.82	4.64	4.50	4.30	4.16	3.78	3.36
	18.63	12.97	10.80	9.63	8.89	8.38	8.00	7.71	7.29	7.00	6.25	5.42

* The entries in this row are quoted to the nearest hundred.

Table 3—*continued*—

v_2 \ v_1	1	2	3	4	5	6	7	8	10	12	24	∞
13	4.67	3.81	3.41	3.18	3.03	2.92	2.83	2.77	2.67	2.60	2.42	2.21
	(6.41)	(4.97)	(4.35)	(4.00)	(3.77)	(3.60)	(3.48)	(3.39)	(3.25)	(3.15)	(2.89)	(2.60)
	9.07	6.70	5.74	5.21	4.86	4.62	4.44	4.30	4.10	3.96	3.59	3.17
	17.82	12.31	10.21	9.07	8.35	7.86	7.49	7.21	6.80	6.52	5.78	4.97
14	4.60	3.74	3.34	3.11	2.96	2.85	2.76	2.70	2.60	2.53	2.35	2.13
	(6.30)	(4.86)	(4.24)	(3.89)	(3.66)	(3.50)	(3.38)	(3.29)	(3.15)	(3.05)	(2.79)	(2.49)
	8.86	6.51	5.56	5.04	4.70	4.46	4.28	4.14	3.94	3.80	3.43	3.00
	17.14	11.78	9.73	8.62	7.92	7.44	7.08	6.80	6.40	6.13	5.41	4.60
16	4.49	3.63	3.24	3.01	2.85	2.74	2.66	2.59	2.49	2.42	2.24	2.01
	(6.12)	(4.69)	(4.08)	(3.73)	(3.50)	(3.34)	(3.22)	(3.12)	(2.99)	(2.89)	(2.63)	(2.32)
	8.53	6.23	5.29	4.77	4.44	4.20	4.03	3.89	3.69	3.55	3.18	2.75
	16.12	10.97	9.01	7.94	7.27	6.80	6.46	6.19	5.81	5.55	4.85	4.06
18	4.41	3.55	3.16	2.93	2.77	2.66	2.58	2.51	2.41	2.34	2.15	1.92
	(5.98)	(4.56)	(3.95)	(3.61)	(3.38)	(3.22)	(3.10)	(3.01)	(2.87)	(2.77)	(2.50)	(2.19)
	8.29	6.01	5.09	4.58	4.25	4.01	3.84	3.71	3.51	3.37	3.00	2.57
	15.38	10.39	8.49	7.46	6.81	6.35	6.02	5.76	5.39	5.13	4.45	3.67
20	4.35	3.49	3.10	2.87	2.71	2.60	2.51	2.45	2.35	2.28	2.08	1.84
	(5.87)	(4.46)	(3.66)	(3.51)	(3.29)	(3.13)	(3.01)	(2.91)	(2.77)	(2.68)	(2.41)	(2.09)
	8.10	5.85	4.94	4.43	4.10	3.87	3.70	3.56	3.37	3.23	2.86	2.42
	14.82	9.95	8.10	7.10	6.46	6.02	5.69	5.44	5.08	4.82	4.15	3.38
22	4.30	3.44	3.05	2.82	2.66	2.55	2.46	2.40	2.30	2.23	2.03	1.78
	(5.79)	(4.38)	(3.78)	(3.44)	(3.22)	(3.05)	(2.93)	(2.84)	(2.70)	(2.60)	(2.33)	(2.00)
	7.95	5.72	4.82	4.31	3.99	3.76	3.59	3.45	3.26	3.12	2.75	2.31
	14.38	9.61	7.80	6.81	6.19	5.76	5.44	5.19	4.83	4.58	3.92	3.15
24	4.26	3.40	3.01	2.78	2.62	2.51	2.42	2.36	2.25	2.18	1.98	1.73
	(5.72)	(4.32)	(3.72)	(3.38)	(3.15)	(2.99)	(2.87)	(2.78)	(2.64)	(2.54)	(2.27)	(1.94)

ν_2												
(24)	7.82	5.61	4.72	4.22	3.90	3.67	3.50	3.36	3.17	3.03	2.66	2.21
	14.03	9.34	7.55	6.59	5.98	5.55	5.23	4.99	4.64	4.39	3.74	2.97
26	4.23	3.37	2.98	2.74	2.59	2.47	2.39	2.32	2.22	2.15	1.95	1.69
	(5.66)	(4.27)	(3.67)	(3.33)	(3.10)	(2.94)	(2.82)	(2.73)	(2.59)	(2.49)	(2.22)	(1.88)
	7.72	5.53	4.64	4.14	3.82	3.59	3.42	3.29	3.09	2.96	2.58	2.13
	13.74	9.12	7.36	6.41	5.80	5.38	5.07	4.83	4.48	4.24	3.59	2.82
28	4.20	3.34	2.95	2.71	2.56	2.45	2.36	2.29	2.19	2.12	1.91	1.65
	(5.61)	(4.22)	(3.63)	(3.29)	(3.06)	(2.90)	(2.78)	(2.69)	(2.55)	(2.45)	(2.17)	(1.83)
	7.64	5.45	4.57	4.07	3.75	3.53	3.36	3.23	3.02	2.90	2.52	2.06
	13.50	8.93	7.19	6.25	5.66	5.24	4.93	4.69	4.35	4.11	3.46	2.69
30	4.17	3.32	2.92	2.69	2.53	2.42	2.33	2.27	2.16	2.09	1.89	1.62
	(5.57)	(4.18)	(3.59)	(3.25)	(3.03)	(2.87)	(2.75)	(2.65)	(2.51)	(2.41)	(2.14)	(1.79)
	7.56	5.39	4.51	4.02	3.70	3.47	3.30	3.17	2.98	2.84	2.47	2.01
	13.29	8.77	7.05	6.12	5.53	5.12	4.82	4.58	4.24	4.00	3.36	2.59
40	4.08	3.23	2.84	2.61	2.45	2.34	2.25	2.18	2.08	2.00	1.79	1.51
	(5.42)	(4.05)	(3.46)	(3.13)	(2.90)	(2.74)	(2.62)	(2.53)	(2.39)	(2.29)	(2.01)	(1.64)
	7.31	5.18	4.31	3.83	3.51	3.29	3.12	2.99	2.80	2.66	2.29	1.80
	12.61	8.25	6.59	5.70	5.13	4.73	4.44	4.21	3.87	3.64	3.01	2.23
60	4.00	3.15	2.76	2.53	2.37	2.25	2.17	2.10	1.99	1.92	1.70	1.39
	(5.29)	(3.93)	(3.34)	(3.01)	(2.79)	(2.63)	(2.51)	(2.41)	(2.27)	(2.17)	(1.88)	(1.48)
	7.08	4.98	4.13	3.65	3.34	3.12	2.95	2.82	2.63	2.50	2.12	1.60
	11.97	7.77	6.17	5.31	4.76	4.37	4.09	3.86	3.54	3.32	2.69	1.89
120	3.92	3.07	2.68	2.45	2.29	2.18	2.09	2.02	1.91	1.83	1.61	1.25
	(5.15)	(3.80)	(3.23)	(2.89)	(2.67)	(2.52)	(2.39)	(2.30)	(2.16)	(2.05)	(1.76)	(1.31)
	6.85	4.79	3.95	3.48	3.17	2.96	2.79	2.66	2.47	2.34	1.95	1.38
	11.38	7.32	5.78	4.95	4.42	4.04	3.77	3.55	3.24	3.02	2.40	1.54
∞	3.84	3.00	2.60	2.37	2.21	2.10	2.01	1.94	1.83	1.75	1.52	1.00
	(5.02)	(3.69)	(3.12)	(2.79)	(2.57)	(2.41)	(2.29)	(2.19)	(2.05)	(1.94)	(1.64)	(1.00)
	6.63	4.61	3.78	3.32	3.02	2.80	2.64	2.51	2.32	2.18	1.79	1.00
	10.83	6.91	5.42	4.62	4.10	3.74	3.47	3.27	2.96	2.74	2.13	1.00

Table 4 Random Numbers

45401	62461	50967	87735	82928	50786	11389	03176	48256	88037
26826	54804	24648	41216	89865	28253	08184	30862	61336	89085
63538	64055	98617	17613	57909	33379	29728	95100	13932	44590
79947	02569	08244	18689	11205	13972	15070	22128	57088	17244
72920	97926	69219	67134	72167	01874	63273	63968	63280	94511
08983	69308	27434	34030	19724	84867	13437	41888	58429	51006
92213	22948	39653	91250	33603	70868	33260	57371	09739	99722
64404	27472	68210	03965	74328	77525	82202	88783	08298	05635
29492	64421	05146	66882	52055	19164	92448	57171	39570	75468
48836	34572	42212	07297	60086	96103	20768	58209	16720	06831
65103	83386	23435	54862	15207	35176	50043	13078	63043	81663
14928	63673	46310	40393	74015	78327	48273	11369	46787	47547
17320	05190	00959	23011	44816	08798	52025	17126	78349	42701
61908	72452	82184	83102	53173	06324	33810	74449	88474	55682
95401	80604	82275	37662	25855	33435	95590	97024	96017	10616
52634	43184	66786	52993	13598	28701	17037	15370	05670	44101
87647	43375	11591	54144	70952	00099	44338	55946	13039	38872
87311	14467	99584	05806	11139	28723	65906	42629	86691	32892
15468	44133	28296	55882	76769	14967	98844	30662	41854	49103
19831	88447	68541	50199	44191	42088	10085	22048	96208	21720
54875	99367	47975	89304	53164	37160	01608	15044	76663	67984
39309	83722	85534	75713	66464	73641	47358	83083	33706	68584
36307	79998	85619	38744	00161	73095	72217	46003	20905	61258
19785	02665	56217	77023	78145	60302	89461	75695	62884	28437
48046	98311	53723	41812	26107	12453	55259	18672	80652	69379
40284	92667	93456	25953	22101	94213	03398	12042	62949	67179
92331	26429	94878	00672	29375	60841	50921	67461	66373	29181
75676	47071	72026	45998	28763	15974	38596	75264	16640	68654
74574	60627	08378	43812	11122	87170	20203	77833	67130	68160
22287	94307	35583	88861	30677	04739	54049	60729	04424	60791
15892	99839	69084	83687	67622	99647	65300	95287	92492	45227
37768	26807	39001	78137	37179	19118	90404	66368	80929	65524
82599	29430	45132	37705	20232	65464	38092	24979	49339	86769
19522	80182	25037	76658	04009	24453	23844	69457	62563	91661
27003	20351	20481	99122	65145	92717	38585	82685	30492	41708
10342	60386	79565	08881	28603	86265	85606	59029	96024	50368
96326	18970	85247	16635	06888	69528	00749	82391	40662	45429
31432	03207	13131	22557	65711	88697	43994	32459	80508	71223
31096	42218	92381	81399	27246	92383	73880	52896	02687	59845
26662	93810	15939	56242	22579	68349	38246	78669	78086	57390
29107	48552	52927	22914	61975	31803	70667	84134	92351	79352
03248	77913	04217	59880	46936	09474	46534	92956	88673	34579
80285	66732	17926	52432	94159	01676	21526	27024	31210	94871
06940	47541	79257	46655	43029	67782	70425	55321	47229	96833
02333	32216	87493	98501	38144	73523	60107	42610	17448	09191
56073	22423	30598	98780	79787	27216	35994	72271	81170	91961
96890	25135	06419	34872	70015	61011	96736	19611	50482	86900
84408	25692	73922	11442	23355	16796	24587	12418	36975	02017
58968	22666	84199	27337	40318	37990	83263	09161	37336	59827

Answers to Problems

Chapter
4. 1. 680, 8, 1, 210, 9,880, 1.
   2. 177,100.
   3. 2,002.
   4. 73,815.
   5. 18, 160, 1476, 108, 134, 16, -38.

Chapter
5. 1. £38,964, £36,400, all nine are modes, £31,400, £7,912, £8,550, £10,526, a little right skewed.
   2. 214.0465, 153, 95, 98,

$$n = 43$$

	153	
62		320
8		859

   851,258,189.4765.
   3. Estimates of 514 and 433, 515.1, 134.4953.
   4. D   5.46875   44.8393.
      E   27.26   7.4591.
   5. 1. 1.7193, 1.65.
      2. 0.6141, 0.85 − 1.94, 1.09.

Chapter
6. 1. 88%, 12%.
   2. 0.312.
   3. 45/80.
   4. 0.18, 0.123, 0.303, NO.
   5. 21/45, 7/24, 1/6, 1/12.
   6. 0.8, NO.
   7. 7/55, 32/45.

Chapter
7. 1. 0.8, 0.778, 2.4, 0.917,
      (d) £67.20, £25.68,
      (e) £6.20, £25.68.

2. (a) b: 1   2   3   4   5

$P(B = b)$:   0.19   0.46   0.28   0.06   0.01

(b) b: 1   2   3   4   5

$P(B = b/T = 1)$:   0.222   0.520   0.257   0.001   0.0

(c) 0.5844

(d) r: 1 $1\frac{1}{3}$ $1\frac{1}{2}$ $1\frac{2}{3}$

$P(R = r)$:   0.205   0.001   0.06   0.004

(d) r: 2 $2\frac{1}{2}$ 3   4

$P(R = r)$:   0.503   0.006   0.22   0.001

    0.227

(e) d: 0   1   2   3   ,   0,289

$P(D = d)$:   0.205   0.506   0.282   0.007

(f) 1.091, 0.5067

Chapter

8. 1. 0.1715.
   2. 0.1493, 0.0628, 7.5, 0.2549.
   4. 0.8159.
   5. 0.1225, 0.0992, 0.001518, 0.1236.
   6. 0.4066, 0.3659, 0.9371.

Chapter

9. 1. 0.2358, 0.695, 0.7257, 0.0918, 0.1816, 0.2463, 0.5169.
   2. 1.08, −0.33, −1.31, 0.78.
   3. 0.0228, 0.2277, 0.567.
   4. 11.12, 7.12, −1.52.
   5. 0.02275, 0.584.
   6. 0.0808, 0.2184, 872 days.
   7. 0.0548.
   8. $28.9\,\mathrm{Nmm^{-2}}$, $29.74\,\mathrm{Nmm^{-2}}$.
   9. 0.479°.
   10. 0.8064, 0.6514.
   11. 0.3594.
   12. £153.72, £29.40, 0.1271, 0.5477.
   13. 12.84 minutes, 13 pumps.

Chapter

11. 1. $N(8.5, 0.10375)$, 0.0606, 118, 1196.
   2. 0.0375, 214.
   3. 0.0287, 0.4714, 0.4714, 812.
   4. 0.3034 years.
   5. £13.98$\frac{1}{2}$.

Chapter

12. 1. (24.613, 29.707).
   2. (129.715, 152.285).

3. (£3,442.97, £4,120.03).
4. (0.584, 0.723).
5. (0.201, 0.351).
6. (0.565, 0.695).
7. (5,280.48, 33,310.75), (£72.67, £182.51).
8. (£251.67, £408.11).
9. (−0.772, −0.348).
10. (−0.052, 0.541).
11. (−0.107 hours, 3.493 hours).
12. (£25.47, £40.54), Yes—matched populations.
13. (−0.0327, 0.1361).
14. (0.0750, 0.3036).
15. (0.3646, 4.7595).
16. (0.9159, 3.5109).

Chapter
13. 1. Two-tailed: reject if $\bar{X}$ is not between 10.775 and 17.545.
   One-tailed: reject if $\bar{X}$ is greater than 17.001.
2. Reject if $\bar{X} < £355.15\frac{1}{2}$. Hence, reject H_0 and conclude
   H_1 that costs have been reduced.
   0.3372, 0.8849, 0.9783.
3. Reject H_0 if $p > 0.55427$. $p = 0.63$. Hence, reject H_0 and
   conclude H_1 that π is significantly greater than 0.5.
4. Reject H_0 if $p > 0.59497$. Need 45.
5. Reject H_0 if $S^2 > 367.886$.
   i.e. reject H_0 if $S > 19.18$.
   Fail to reject H_0.
6. Reject H_0 if S is not between 5.18 and 9.302. $S = 4.31$.
   Hence, reject H_0 and conclude H_1 that there is a difference in standard deviation.
7. Reject $H_0: \sigma_1^2 = \sigma_2^2$ if S^2L/S^2S (larger divided by smaller)
   > 2.01. $S^2L/S^2S = 2.19$.
   Hence, reject H_0 and conclude H_1 that there is a difference.
8. Reject H_0 if $r^* > 0.3051$. $r^* = 0.28226$. Hence, fail to
   reject H_0.
9. Reject H_0 if $|r^*| > 0.1559$.
   $r^* = -0.1923$. Hence, reject H_0 and conclude H_1 that the
   correlation is significantly different from zero.
10. If using pooled variance, reject H_0 if $\bar{X}_1 - \bar{X}_2 > 0.6643$.
   If using separate variances, reject H_0 if $\bar{X}_1 - \bar{X}_2 > 0.7139$.
   $\bar{X}_1 - \bar{X}_2 = 0.8$. Hence, with either approach, we reject
   H_0 and conclude H_1 that workers in group A work, on
   average, more overtime than workers in group B.

Chapter
14. No answers are given as the answers depend upon the models used.

Chapter
15. 1. Exp = 32.993 + 0.55096 Inc.
    slope is significantly positive.
    (£112.69, £115.38).
    2. Exp = 76,453 − 3,408 Int $R^2 = 31.7\%$
    p.c. Exp = 3,050 − 61.57 Int $R^2 = 7.9\%$
    Is the latter model of any use with such a low R^2?
    3. Price = 1.5557 + 0.23925 Income.
    Intercept is significantly non-zero.
    (£4,695.65, £5,545.40).
    Income = −5.859 + 4.0774 Price.
    Income is unlikely to be known exactly. Therefore, we cannot use income as an explanatory variable. Therefore, the latter model looks more likely to be valid. (However, while price may be known exactly, it is subject to market forces and may also be altered slightly to entice customers of high or of low income.)

Chapter
17. No answers are given as the answers depend upon the index used.

Index